PROTEIN–NUCLEIC ACID INTERACTION

TOPICS IN MOLECULAR AND STRUCTURAL BIOLOGY

Series Editors

Stephen Neidle
Institute of Cancer Research
Sutton, Surrey, UK

Watson Fuller
Department of Physics
University of Keele, UK

Protein–Nucleic Acid Interaction
Edited by Wolfram Saenger and Udo Heinemann

Calcified Tissue
Edited by David Hukins

Oligodeoxynucleotides
Edited by Jack S. Cohen

Molecular Mechanisms in Muscular Contraction
Edited by John N. Squire

PROTEIN–NUCLEIC ACID INTERACTION

Edited by

WOLFRAM SAENGER and UDO HEINEMANN

Institut für Kristallographie
Freie Universität Berlin,
West Germany

CRC Press, Inc.
Boca Raton, Florida

First published 1989

Published in the USA, its dependencies, and Canada by
CRC Press, Inc.
2000 Corporate Blvd., N.W.
Boca Raton, FL 33431, U.S.A.

Typeset by Wearside Tradespools Ltd
Fulwell, Sunderland

Printed in Great Britain

Catalog #Z7113
ISBN 0–8493–7113–9
ISSN 0265–4377

Contents

The Contributors

Jan Antosiewicz
Max-Planck-Institut für
biophysikalische Chemie
3400 Göttingen
West Germany

Otto G. Berg
Department of Molecular Biology
Uppsala University Biomedical Centre
Box 590
751 24 Uppsala
Sweden

Rolf Boelens
Department of Chemistry
University of Utrecht
Padualaan 8
3584 CH Utrecht
The Netherlands

Joachim Greipel
Medizinische Hochschule Hannover
Zentrum Biochemie
Abt. Biophysikalische Chemie
3000 Hannover
West Germany

Ulrich Hahn
Institut für Kristallographie
Freie Universität Berlin
Takustr. 6
1000 Berlin 33
West Germany

Udo Heinemann
Institut für Kristallographie
Freie Universität Berlin
Takustr. 6
1000 Berlin 33
West Germany

Wolfgang Hillen
Lehrstuhl für Mikrobiologie
Institut für Mikrobiologie und
Biochemie
Staudtstr. 5
8520 Erlangen
West Germany

Peter H. von Hippel
Institute of Molecular Biology
University of Oregon
Eugene
OR 97403
USA

Robert Kaptein
Department of Chemistry
University of Utrecht
Padualaan 8
3584 CH Utrecht
The Netherlands

M. H. J. Koch
European Molecular Biology
Laboratory
c/o DESY
Notkestr. 85
2000 Hamburg 52
West Germany

Rolf M. J. N. Lamerichs
Department of Chemistry
University of Utrecht
Padualaan 8
3584 CH Utrecht
The Netherlands

Günter Maass
Medizinische Hochschule Hannover
Zentrum Biochemie
Abt. Biophysikalische Chemie
3000 Hannover
West Germany

Dietmar Porschke
Max-Planck-Institut für
biophysikalische Chemie
3400 Göttingen
West Germany

Gerald Stubbs
Department of Molecular Biology
Box 1820, Station B
Vanderbilt University
Nashville
TN 37235
USA

Claus Urbanke
Medizinische Hochschule Hannover
Zentrum Biochemie
Abt. Biophysikalische Chemie
3000 Hannover
West Germany

Andreas Wissmann
Lehrstuhl für Mikrobiologie
Institut für Mikrobiologie und
Biochemie
Staudtstr. 5
8520 Erlangen
West Germany

Preface

Protein–nucleic acid interactions have been a focal point of scientific interest over the past three decades. The number of exciting papers that appear annually on this topic is breathtaking. Therefore, it could not be our intention to provide with this volume a comprehensive overview of the field. Instead, we have chosen to bring together contributions describing a wide range of problems in the field of protein–nucleic acid interactions investigated by a variety of techniques.

This volume begins with a chapter on DNA–protein interactions in the regulation of gene expression, which may serve as a convenient entry point into the book. The remaining chapters may be read in any order, since each is devoted to a selected model system. These cover several orders of magnitude in size, going from small proteins like Lac repressor headpiece and RNase T1 and their cognate nucleic acid fragments to chromatin, thus reflecting the amazing diversity of problems even in a fairly well-defined scientific field such as the present. Our main emphasis in putting together this volume has been to include all important techniques currently used in studying protein–nucleic acid interactions. We believe that this aim has been reached within the limits imposed by the size of the book. It goes without saying, and becomes perfectly clear when reading this volume, that only the interplay of the various techniques may be expected to bring about new developments in this field.

Without the very careful work of the contributors, this volume would not have been possible. We would like to thank all authors for their diligence, Stephen Neidle and Watson Fuller, the series editors, for the impetus to assemble this book and Harry Holt and David Grist of The Macmillan Press for their patience and good co-operation.

Berlin, 1989   W.S.
U.H.

1
DNA–protein interactions in the regulation of gene expression

Peter H. von Hippel and Otto G. Berg

INTRODUCTION

In this paper we describe a number of approaches to the nature and specificity of DNA–protein interactions involved in the regulation of gene expression at the transcriptional level. We discuss primarily the binding of 'single-specific-site' regulatory proteins to DNA targets. This involves consideration of several aspects of the specificity of DNA–protein interactions, including: (1) the combinatorial specification of the number of base pairs required to define a unique binding site in a genome of given size; (2) the structure of DNA and protein binding sites, including structural complementarity and steric aspects; (3) the energetics of the binding interaction, including both specific and non-specific binding; (4) the thermodynamics and kinetics of the overall interaction, as determined by the net binding free energy of specific complex formation and the effects of competing sites; and (5) equilibrium binding selection, which determines the actual level of saturation of the specific (regulatory) target under various environmental conditions. These aspects are all interdependent, and a coherent picture of the specificity of such interactions can be obtained only by considering them all in context. Such an approach has been set forth by us previously, in part, in von Hippel and Berg (1986), and portions of this overview are taken directly from that treatment. In conclusion, we consider how these ideas modulate the evolutionary 'design' of regulatory proteins, as well as the formulation of purification procedures and binding assays for these proteins, and how these approaches may apply *in vivo*.

SINGLE PROTEIN BINDING TO A REGULATORY DNA SITE

Regulation of transcription at the DNA level clearly must involve, at least as an initial step, the binding of the regulatory protein to a DNA target site. This binding may comprise the entire process, as in repressor binding directly to its operator target in competition with RNA polymerase. A regulatory protein may also operate as an activator, changing the vicinal affinity, and thus the activity, of a 'primary' regulatory protein such as RNA polymerase. And finally, of course, the protein may itself *be* a primary regulatory protein, for which binding serves only as the first of a series of sequential functional steps (e.g., RNA polymerase interacting with promoter to form a closed polymerase–promoter complex).

The general problem of DNA–protein interaction specificity is best described in functional terms; i.e., in terms of the degree of effective saturation of a regulatory target site on a particular chromosome. The *lac* operon of *E. coli* provides a useful illustration. This operon occurs once per bacterial genome. Depending on the physiological state of *E. coli*, and thus on its level of replication in proportion to the rate of cell division, the average bacterium may contain one or several (up to 3 or 4) copies of the *lac* operon. Each operon contains one operator site, which serves as the specific binding target for *lac* repressor. In wild-type cells there is an average of ten to thirty copies of the *lac* repressor protein. The central protein–nucleic acid interaction that is thought to define this system is the competitive (with RNA polymerase) binding of *lac* repressor to *lac* operator. The repressor and polymerase binding sites (operator and promoter, respectively) overlap (Dickson *et al.*, 1975); thus, when repressor is bound, the promoter is at least partially occluded and transcription is inhibited. The *lac* operon in wild-type *E. coli* is expressed at $\sim 10^{-3}$ of the induced (constitutive or *lac*$^-$) level. In binding terms this means that the ratio of free to repressor-complexed operator sites *in vivo* is $\sim 10^{-3}$. A detailed thermodynamic analysis of this system has been presented elsewhere (von Hippel *et al.*, 1974; von Hippel, 1979).

The degree of specificity of this interaction can best be appreciated when one realizes that the *in vivo* system contains $\sim 10^7$ DNA binding sites that can, in principle, compete for *lac* repressor (each base pair of the chromosome comprises the beginning of a potential competing binding site). Thus the total concentration of potential DNA binding sites (D_T) greatly exceeds that of repressor molecules (R_T), which in turn exceeds that of operator sites (O_T); i.e., $D_T \gg R_T > O_T$.

The effective specificity of the system is measured in terms of the fractional saturation of the operator site with repressor. Obviously this will relate to the free concentrations of the various species, and will thus depend, in large measure, on ratios of specific to non-specific binding constants.

LEVELS OF SPECIFICITY

Binding site specification

We first consider the system in terms of absolute specificity; i.e., we assume a protein that can *absolutely* (and *only*) discriminate between the four 'information elements' of DNA. These are four canonical nucleotide residues (A, T, G, and C) in single-stranded DNA and the four types of base pairs (A·T, T·A, G·C and C·G) in double-stranded DNA. The latter will be our primary focus here. (An A·T pair can be discriminated from a T·A pair because of the polar (5′→3′) character of the sugar–phosphate backbones of the individual DNA chains.)

A specific binding site is defined as a particular sequence of base pairs. A conditional probability approach (von Hippel, 1979) can be used to determine n, the *minimum* length of a sequence of recognition elements (base pairs) required to specify a site, so that the expected frequency $(P_n \cdot 2N)$ with which that site reappears at random within the genome is less than unity. An example of this approach is presented in Table 1.1. For *E. coli* DNA, this minimal length is ~12 base pairs, assuming a double-stranded sequence within a genome of overall composition A = T = G = C; see Figure 1.1.

Table 1.1   Conditional probability approach to DNA sequence specification

Position indices:		1	2	3	4	5	6	7	8	
	(5′–)	A	T	G	C	G	T	T	C	(–3′)
	(3′–)	T	A	C	G	C	A	A	G	(–5′)
Probabilities in each position		$P_{1,A}$	$P_{2,T}$	$P_{3,G}$	$P_{4,C}$	$P_{5,G}$	$P_{6,T}$	$P_{7,T}$	$P_{8,C}$	

For $N \gg n$, the probability of occurrence of this particular sequence at a specific position i in the genome is:

$$P_n = (P_{1,A})(P_{2,T})(P_{3,G})(P_{4,C})(P_{5,G})(P_{6,T})(P_{7,T})(P_{8,C})$$

For genomes containing equal numbers of A·T, T·A, G·C, and C·G base pairs, $P_A = P_T = P_G = P_C = 0.25$, and the probability of occurrence is:

$$P_n = (P_A)(P_T)^3(P_G)^2(P_C)^2 = (0.25)^8 = 1.526 \times 10^{-5}$$

The expected frequency of random occurrence of this particular sequence *in an entire genome* of N base pairs is:

$$(P_n)(2N) = (1.526 \times 10^{-5})(2 \times 10^7) = 305 \text{ for a genome of } 10^7 \text{ base pairs (potential binding}$$
sites). (Each base pair represents the beginning of a potential binding site; the factor 2 enters because a binding site in double-stranded DNA can be read 5′→3′ along either strand.)

From Figure 1.1, it can be seen that if the site size is smaller than $n \simeq 12$ then one can expect a large number of sites identical in sequence to the specific site to occur at random in the genome. Such sites will serve as competitive traps (sinks) and will decrease the amount of free protein

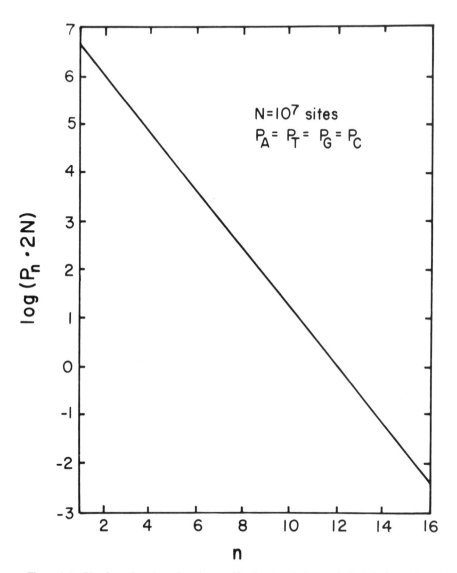

Figure 1.1   Plot (as a function of n, the specific site size, in base pairs) of the logarithm of $P_n \cdot 2N$, the expected frequency of random occurrence of a specific base pair sequence of length n within a genome containing $N = 10^7$ sites, for a DNA in which the individual probabilities of $A \cdot T$, $T \cdot A$, $G \cdot C$ and $C \cdot G$ base pairs are all equal ($P_A = P_T = P_G = P_C$). (Taken from von Hippel, 1979)

available to interact with the functional target site(s). Thus, $n = 12$ corresponds to a minimal specification for a specific site. However, sites differing by only one or a few base pairs will still be very numerous; from the combinatorial possibilities one expects $3n = 36$ sites with one base pair 'wrong', $9n(n-1)/2 = 594$ with two base pairs wrong, etc. Unless the affinity for these partially specific sites decreases very rapidly with increasing number of wrong base pairs, they will compete strongly for the protein through their large numbers. Consequently, depending on the affinity changes, one would expect some degree of over-specification, so that $n > 12$ (von Hippel, 1979; von Hippel and Berg, 1986) and the competition from partially wrong sites can be kept to a reasonable level.

This approach assumes that the overall sequence of the genome can be treated as chemically (though obviously not genetically) random. This assumption is supported (for sequences above the tetranucleotide level) both by comparison of calculated versus observed restriction enzyme cuts in viral genomes of defined size (von Hippel, 1979), and by the frequencies of occurrence of specific sequences tabulated by computer search of genomic libraries. We also assume that every base pair is fully specified (in terms of base pair type). Unspecified loci can, of course, interrupt the overall sequence in defined positions, but these loci will not count toward n. Similarly, specification only at the level of Pu·Py (purine·pyrimidine) (vs. Py·Pu) base pairs can occur; such loci are weighted less in establishing n. (For further details of this approach, see von Hippel, 1979.)

The chemical basis of recognition

Primary sequence recognition mechanisms
The primary molecular mechanism that can unambiguously recognize and discriminate *individual* base pairs in double-stranded DNA is complementary hydrogen bonding. Just as Watson–Crick base–base interactions involve the articulation of complementary hydrogen bond donor and acceptor groups between the central 'faces' of the bases to discriminate 'right' from 'wrong' base-pairing in the template recognition events involved in replication and transcription, so also will the hydrogen bond donor and acceptor groups of amino acid residues of the protein binding site serve primarily to discriminate base pairs by complementary hydrogen bonding with the acceptors and donors of the major and minor grooves of the double helix (Yarus, 1969; von Hippel and McGhee, 1972; Seeman *et al.*, 1976). These functional groups are shown in Figure 1.2 (modified after Seeman *et al.*, 1976) and in a useful 'stick-figure' representation in Figure 1.3 (see Woodbury *et al.*, 1980; Woodbury and von Hippel, 1981; von Hippel *et al.*, 1982; Ohlendorf *et al.*, 1982). A stick figure representation of the sequence recognized by the *EcoRI* restriction and modification

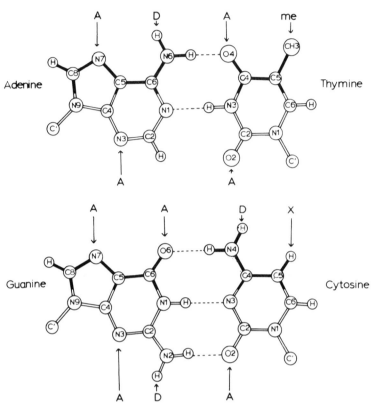

Figure 1.2   Molecular models of an $A \cdot T$ and $G \cdot C$ base pair, showing the functional groups in the major and minor grooves that may be important for protein recognition. A = hydrogen-bond acceptor; D = hydrogen-bond donor; me = the methyl group of thymine; X = the readily modified 5-position of cytosine

enzymes is shown in Figure 1.4. Clearly, *all* these hydrogen donor and acceptor interactions are not utilized in any single protein–DNA acid binding interaction, nor are the functional groups in both the major and the minor grooves used in recognizing any one base pair. Also, the DNA will not always remain in the undistorted double-stranded B conformation. Nevertheless, Figure 1.4 does show the repertoire of possibilities from which a selection of recognition elements *can* be made by a binding protein.

Secondary sequence recognition mechanisms
When the issue is not the absolute identification of a particular sequence of base pairs by a regulatory protein, other recognition mechanisms, at a 'lower level' of specification, can also come into play. Thus DNA 'regions' can be discriminated at the level of strandedness (e.g., single- vs. double-stranded), groove geometry and secondary structure (e.g., B- vs. Z-form

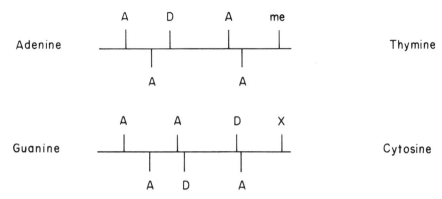

Figure 1.3 Schematic ('stick-figure') representations of the functional groups of an $A \cdot T$ and a $G \cdot C$ base pair that may be important in protein recognition. (Taken from Woodbury and von Hippel, 1981)

DNA), etc., based on differences in protein binding affinity. The origins of these affinity differences can be steric, hydrophobic and/or electrostatic, and will reflect structural consequences of regional differences in base pair composition. Examples include the preferred binding of tetra-alkylammonium ions to $dA \cdot dT$ sequences in the (major) groove of B-DNA (Melchior and von Hippel, 1973), the preferred binding of the antibiotics netropsin and distamycin in the minor grooves of poly($dA \cdot dT$) and of $dA \cdot dT$-rich sequences of B-form DNA (Kopka *et al.*, 1985), the non-specific binding of *E. coli lac* repressor to double-stranded DNA (deHaseth *et al.*, 1977; Revzin and von Hippel, 1977), the preferential binding of phage T4-coded gene 32 protein to single-stranded DNA and RNA (Kowalczykowski *et al.*, 1981) and the differential sensitivities of gene-specific eukaryotic DNA regions to endonucleases (Weintraub and Groudine, 1976). Such interactions obviously provide a measure of regional binding specificity. However, they do not carry enough structural information to provide a mechanism for the primary recognition of specific base pair sequences by proteins in most cases. We expect that primary recognition will be dominated by hydrogen bonding interactions.

Affinity

Accepting the notion that specific target site recognition does indeed involve the 'reading', by a regulatory protein, of a specific array of hydrogen bonding donors and acceptors in the major and minor grooves of the DNA double helix, we next consider the question of quantitative specificity or discrimination, which can also be termed 'the problem of the other sites'.

This problem exists because discrimination between (e.g.) 'right' and 'wrong' base pairs cannot be absolute. Rather, there is some finite level of

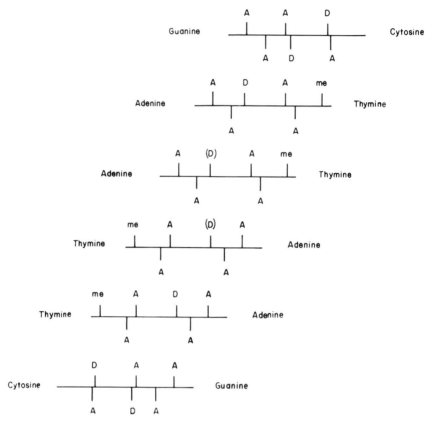

Figure 1.4 Schematic representation of the canonical-recognition base pair sequence for *Eco*RI endonuclease and methylase. All primary functional groups are shown as in Figure 1.3. (D) represents hydrogen-bond donors in this sequence that are subject to methylation by the *Eco*RI modification enzyme. (Taken from Woodbury *et al.*, 1980)

affinity of the protein for the 'correct' site, and some lower (but non-zero) and progressively decreasing affinity for other sites with decreasing degrees of homology with the correct one. To the extent that the great preponderance of 'wrong' sites can compete with the regulatory target for protein, and thus reduce the free protein concentration, the effective affinity of the protein for the correct sites will also be reduced.

This idea is shown schematically in Figure 1.5, which provides a specific, but over-simplified, illustration of the notion that recognition is not absolute. Figure 1.5 shows the (non-cooperative) titration by (e.g.) repressor of a specific (e.g.) operator site, labelled O_1, followed, at higher free protein concentrations, by the titration of a somewhat heterogeneous group of related operator sites located on other genes and labelled 'O_2, O_3, O_4 . . .'. Figure 1.5 points out that this pair of titrations defines a 'window of specificity' in terms of free protein concentration; at very low protein concentration there is no specific binding, because the free protein concentration is below the affinity 'threshold' for site O_1. At high protein concentrations, repression is no longer specific only for O_1. Only at concentrations between the two binding isotherms do we see what we define as *specific* functional binding of this particular regulatory protein.

Representations such as Figure 1.5 also make it clear that while the window of specificity spans a finite protein concentration range, its position and size will depend on the affinity of the protein for sites O_1 *and* sites O_2, O_3, etc. Mutations in either the protein or the nucleic acid sites, or changes in the composition of the cellular environment, can narrow the specificity window, widen it, or obliterate it all together. In addition, binding of other proteins manifesting positive co-operativity to an extended operator site can both sharpen the specific titration and strengthen the effective binding (see Figure 1.5), thus also widening the window of specificity.

Sequence-specific binding free energy
We can attempt to estimate the favourable binding free energy expected per correctly positioned hydrogen bond donor–acceptor pair between protein and nucleic acid from first principles. Since the functional groups of

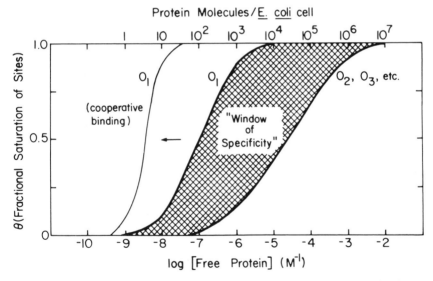

Figure 1.5   Titration of a specific operator site O_1 and of pseudospecific sites $O_2, O_3, O_4 \ldots$ with repressor. Fractional saturation (θ) is plotted as a function of the logarithm of the *free* protein (repressor) concentration. The numbers at the top of the graph indicate the number of protein molecules per *E. coli* cell (assumed volume $= 10^{-15}$ litres) that these concentrations represent. The cross-hatched area between the two titration curves defines the 'window of specificity' for this binding system. Co-operative binding of repressor to O_1 can strengthen the effective binding (shift it to the left) *and* sharpen the binding isotherm (see text)

both the protein and the nucleic acid binding sites will be involved in hydrogen bonding with water molecules when the complex is dissociated, for illustrative purposes we assume an average (*differential*) contribution of ~−0.5 kcal/mole for each correctly formed protein to nucleic acid hydrogen bond. Assuming an average of one to two hydrogen-bonded recognition events per base pair, this gives us a range of favourable specific binding free energies (for a protein with a recognition site size (n) of 12 base pairs) of ~−6 to ~−12 kcal/mole protein bound. (We note (see Seeman *et al.*, 1976) that two protein–nucleic acid hydrogen bonds will generally be required to recognize one base pair *in isolation*. However, one specifically positioned hydrogen bond may suffice in the context of a base pair 'stack', since the hydrogen bonding groups of both the protein and nucleic acid are held in fixed positions relative to the rest of the nucleic acid sequence and protein binding site in such structures.)

These are not large numbers, and it is important to recognize that much more favourable free energy is likely to be lost per mispaired position than is gained per proper recognition event. This follows because the interaction of a protein 'recognition' functional group with an 'incorrect' base pair can result in the total loss of at least one hydrogen bonding interaction; i.e., a protein hydrogen bond donor may end up 'facing' a nucleic acid donor, or an acceptor may be 'buried' facing an acceptor. In either case, at least one hydrogen bond that was broken in removing the protein and nucleic acid donor (or acceptor) groups from contact with the solvent is not replaced, and an unfavourable contribution of as much as +5 kcal/mole may be added to the binding free energy, unless the protein–DNA complex can adjust its overall conformation somewhat to minimize this problem. (Recently, Tronrud *et al.* (1987) have obtained a direct measure of this maximum destabilization effect by comparing two enzyme–substrate complexes, in one of which a proper hydrogen bond is formed, while in the other a functional group of the (otherwise identical) substrate has been altered from a hydrogen bond donor to an acceptor, thus leaving two acceptor groups 'facing' one another. Crystallographic analyses show that the two ES complexes involved are entirely isomorphous, and the locus of the putative hydrogen bond is entirely shielded from water. Binding constant measurements show that the stability of the two complexes differs by ~4 kcal/mole, close to the value expected for the uncompensated loss of one hydrogen bond.)

This phenomenon illustrates the principle that generally applies to recognition interactions based on hydrogen bond donor–acceptor complementarity in water; i.e., *correct* acceptor–donor interactions may not add much to the stability of the complex, but *incorrect* hydrogen bond complementarities are markedly destabilizing. Thus, differential specificity of this type can usually be attributed to the *unfavourable* effects of incorrect contacts.

Non-sequence-specific binding free energy

If, as discussed above, the main determinants of specificity are the unfavourable contributions of 'wrong' base pairs, specific binding will also require a large non-specific contribution to the binding free energy to achieve sufficient binding affinity. Such non-specific interactions usually involve a large electrostatic component, due mostly to the displacement of condensed counterions from DNA phosphate groups by positively charged protein side-chains (see Record *et al.*, 1976). For example, for *lac* repressor binding to the operator site, we estimate a total standard free energy of binding of approximately −17 kcal/mole under physiological salt concentrations. This interaction involves the formation of 7–8 non-specific charge–charge interactions between protein and DNA (Record *et al.*, 1977; Winter and von Hippel, 1981) and numerous base-pair-specific recognition interactions (Goeddel *et al.*, 1978). The binding of *lac* repressor to non-specific DNA involves charge–charge interactions, and no base pair specific interactions (deHaseth *et al.*, 1977; Revzin and von Hippel, 1977), resulting in a standard free energy of binding of approximately −7 kcal/mole under the same conditions.

Conformational change of the regulatory protein to a totally non-specific binding mode

The above discussion suggests that at the rate at which the specific binding free energy decreases with misplaced (non-complementary) hydrogen-bonding contacts, more than 3 to 5 'incorrect' base-pairs in the *lac* operator sequence may result in complete dissociation of a repressor–pseudo-operator complex. Instead, we find that under these conditions *lac* repressor 'isomerizes' to a binding mode in which the interaction free energy is totally electrostatic and involves no sequence-dependent components. The same behaviour may also characterize *E. coli* RNA polymerase (deHaseth *et al.*, 1978) and T4-coded DNA polymerase (Fairfield *et al.*, 1983). One of the advantages of this non-specific binding mode for *lac* repressor (and perhaps for other genome-regulatory proteins as well) may involve the ability of the protein to 'slide' over the surface of the DNA molecule in a one-dimensional diffusion process in this binding mode, thus facilitating translocation to the regulatory target site (Winter *et al.*, 1981; Berg *et al.*, 1982).

It is important to realize that such conformational lability in the protein will decrease specificity, since it strengthens the binding to non-specific sites. The protein can also have different specific binding modes that are induced by different specific sequences (Mossing and Record, 1985). Similarly, such lability will also decrease specificity, since it increases the number of possible base pair sequences to which the protein can bind.

Distortions of the protein and/or the DNA target site

In concluding these remarks on affinity, it is important to stress that neither the protein nor the DNA sites involved in binding are totally rigid. Thus, both partners in complex formation can (and will) distort to optimize sterically sensitive binding interactions, within the limits of energetically available conformations. We have already seen an example of this, in the isomerization of *lac* repressor to a totally electrostatically bound, non-sequence-dependent binding mode in the presence of an excess of un-favourable hydrogen bonding interactions. DNA sites can also change their local conformations. Thus, for example, hydrogen bonding of DNA base pairs to a potentially complementary protein matrix could be either improved or degraded by a local conformational change of the DNA that modifies the relative positions, directions and exposures of the hydrogen bonding functional groups located in the grooves of the DNA double helix.

Since both the conformation and flexibility of DNA are dependent on the sequence, such distortions in the DNA can also contribute to the specificity of the protein interactions. Certain DNA sequences may be easier to distort to fit the binding site of the protein and some base pair choices can therefore be important for specificity even if they do not contribute any specific interactions. Such effects have been observed or postulated for the binding of λ 434 repressor to its operator (Anderson *et al.*, 1987) and also for the binding of the cyclic AMP receptor protein to its recognition sites (Liu-Johnson *et al.*, 1986). In this way, sequence-dependent DNA flexibility can increase binding specificity. However, to be effective, this mechanism also requires a 'stiff' protein; otherwise, the conformational distortions will take place in the protein rather than in the DNA. Thus specificity can be enhanced by DNA distortions and reduced by protein distortions.

It is important to emphasize that such distortions from optimal solution conformations are associated with a thermodynamic cost. The free energy required to maintain the optimal (distorted) binding conformation of either partner must be subtracted from the favourable free energy of the binding interaction. Beyond a certain point, this thermodynamic cost exceeds the free energy gained as a consequence of the interaction, and complex formation no longer occurs.

Equilibrium selection

How are binding sequences designed to provide sufficient specificity *in vivo*? In the example of repressor–operator binding, the effective (func-tional) selection is determined by the fractional saturation of the operator site(s). And this, in turn, is controlled both by the affinities discussed above *and* by 'pseudo-operator' and non-specific DNA binding sites.

These overall competitive binding reactions can best be described by a coupled equilibrium model, which describes the binding of (e.g.) repressor to the entire distribution of binding sites that show some affinity for it, and ultimately control the free repressor concentration and thus the final level of repression manifested by the system. Clearly, this is not just a 'nuisance'; as shown earlier (von Hippel *et al.*, 1974; von Hippel, 1979), the overall level of repression of the *lac* system, as well as the ability of inducer to derepress the system, and the effects of various repressor and operator mutations, can only be understood within the context of such coupled equilibrium calculations. The *quantitative* treatment of such systems, including the handling of binding to various classes of 'pseudo-specific' sites (i.e. sites that differ from the specific site in only a few base-pairing positions and to which the regulatory protein will bind with approximately the same conformation that it uses to bind to the specific site) and completely non-specific sites, is described in von Hippel and Berg (1986) and in Berg and von Hippel (1987, 1988).

Actually, the existence of the non-specific and pseudo-specific sites contributes the major evolutionary selection pressure for the specificity of the functional sites; the specificity is important only relative to the competitive binding reactions. The various selection pressures acting on the functional sites will show up in the statistics of base-pair choice in these sites, and in fact one can use these statistics to estimate various interaction parameters and also the extent of pseudosite binding in the living cell (Berg and von Hippel, 1987, 1988).

CONCLUSIONS, APPLICATIONS, FURTHER DEVELOPMENTS AND MORE COMPLICATED PROBLEMS

In this paper, we have summarized a series of approaches to the specificity of protein–nucleic acid interactions involved in the regulation (at the DNA level) of biological function. The ultimate manifestation of regulatory specificity must be the degree to which a specific biological process (e.g., the transcription of a particular gene or the translation into a particular protein) is 'turned up' or 'turned down' as a consequence of a specific DNA–protein interaction. Even though a given process of gene expression may reflect the end-product of a lengthy series of pre-equilibrium, steady-state or kinetically controlled steps, over an appreciable range of rates and concentrations most regulatory processes will reflect directly the equilibrium extent of specific DNA target saturation by the relevant binding protein(s), and it is this level of specificity with which we are concerned here.

Nomenclature

In order to discuss this problem with precision, we have attempted to differentiate the various levels of specificity with an appropriate terminology. Thus specification (or 'information') refers to the length (in base pairs) of the sequence actually involved in specifying the target binding site (the term 'information content' will be reserved for its precise statistical meaning; see Schneider *et al.*, 1986; Berg and von Hippel, 1987, 1988). Recognition is defined by the physico-chemical mechanisms that actually control the specificity of the interactions. Discrimination (or selectivity) refers to the thermodynamics of the interactions involved, and is determined by the differences in affinity of the protein for the various DNA targets over which the protein is distributed. Thus, the discrimination ratio for pairs of binding sites is a ratio of binding constants. Finally, selection (in the equilibrium case), or the final level of biological expression (which reflects regulatory site saturation), is determined by the effective binding relation for the whole system of proteins and DNA binding sites. This hierarchy of specificity considerations makes it possible to examine a number of issues related to functional specificity.

Evolutionary 'design' of regulatory proteins

As indicated above, in order to saturate a regulatory DNA site adequately, the absolute binding affinity of the protein must be high enough to permit fairly complete titration of the site at the level of free protein concentration present in the cell. Operating under conditions in which the concentration of free protein is effectively determined by the concentrations and affinities for protein of the 'other' DNA binding sites of the genome, the *minimum* specific binding constant required depends on the various selectivity ratios involved and can be calculated using equation (5) of von Hippel and Berg (1986). The *maximum* value of the specific binding constant is also constrained, since the dissociation time of the protein must be set below the cell cycle times needed to achieve (e.g.) DNA replication, organelle duplication and cell division (though we note that the cell may devise special allosteric mechanisms, such as inducer binding, to lower the binding constant – and thus raise the dissociation rate – into a biologically acceptable range). Regulatory protein and DNA binding site 'design' may cope with such problems in various ways.

(1) Proteins may be composed of (loosely or tightly associated) identical subunits, with two or more subunits binding in tandem to regulatory DNA binding sites of repeating base-pair sequence (examples include *E. coli lac* repressor, lambda CI or *cro* repressor, etc.). The advantages of this procedure include the fact that genetically coded protein units can be small and still can recognize a large site, i.e., a site specified by the totality of the tandem sequences. Additional levels of regulation of net binding affinity can be introduced by increasing or decreasing the inter-protein-subunit binding affinity, and thus the effective (at the subunit level) degree of protein binding co-operativity.

(2) The overall protein may also be designed to bind with intermolecular

co-operativity to regulatory sites, i.e., the affinity of the protein for a subsite of (e.g.) an operator can be increased (by favourable protein–protein interactions or by local DNA-lattice deformation) by the binding of another protein of the same type (homoprotein co-operativity) or of a different type (heteroprotein co-operativity) to a contiguous site. (For a treatment of the lambda phage lytic lysogenic regulatory switching system, which clearly operates this way, see Ptashne *et al.* (1980). The general principles of such systems, in the context of the ribosome assembly problem, have recently been formulated by von Hippel and Fairfield, 1985, and Fairfield and von Hippel, in preparation.)

(3) The possibility of at least two protein binding conformations also introduces an additional regulatory variable. Thus, as indicated in the preceding section, a protein can bind specifically to the regulatory site (and to pseudosites differing from the canonical sequence in only a very few base-pair positions), and can assume another (totally non-specifically binding) conformation that becomes favourable for binding when the number of incorrect (and thermodynamically unfavourable) base pairs exceeds some number j (see von Hippel and Berg, 1986). This additional degree of conformational freedom at the protein level permits the establishment of a constant discrimination ratio for the majority of the available sites on the genome. Furthermore, if this general non-specific affinity is totally electrostatic, and if the salt dependence of the specific and non-specific binding constants differ, this discrimination ratio can be manipulated by small changes in the ionic environment. In addition, of course, such totally electrostatic non-specific affinity of the regulatory protein for the overall genome may permit the facilitated location of regulatory sites (see above).

(4) Small-molecule ligands that can bind to the regulatory protein, and alter its affinity for the regulatory DNA target (e.g., inducers, as in the *lac* system), provide another dimension of regulatory control, and can be included by expanding the system of components in the overall coupled equilibrium system (see von Hippel *et al.*, 1974; von Hippel, 1979; O'Gorman *et al.*, 1980).

Non-specific DNA binding and the formulation of purification procedures and binding assays for regulatory proteins

The existence of non-specific DNA binding of regulatory proteins considerably complicates the design of experiments to purify and assay these proteins, especially if the specific and non-specific binding affinities vary differently with salt concentration, because then the selectivity ratios of the protein become a function of this variable. Most protein purifications and *in vitro* assays are conducted at low salt concentrations, generally because protein concentrations are limited and tighter binding is achieved at low salt. The consequence of this, however, is often that the competition of non-specific DNA sites for protein (which is generally very intense at low salt) completely 'swamps' binding to the putative (and often correctly identified) specific sites, and renders worthless the classical purification paradigm of using specific DNA sites (e.g., in a column) to isolate the regulatory protein from an impure extract. In the same way, an excess of non-specific DNA, attached to an isolated DNA fragment containing the site of interest, can completely suppress specific binding under conditions where the selectivity ratio for specific binding is too low. It is clearly important to take these considerations into account in formulating purification and assay procedures, since, if one does not set up the experiment within a 'window' of salt concentration at which specific site selection is possible, the experiment is doomed to fail. The 'take-home' lesson may be that salt concentration must be a variable in the design of a purification or assay procedure for any protein for which the specific or non-specific binding parameters depend on this factor.

Binding selectivity and affinities *in vivo*

In addition to developing insight into protein–DNA interaction mechanisms and possibilities, the final objective of studies of this sort must be to understand the situation *in vivo* in both prokaryotic and eukaryotic cells. To define *in vivo* levels of selection, it is necessary to extrapolate discrimination ratios established by *in vitro* experiments to *in vivo* conditions. This requires that the ionic environment, genome size, concentration of total and of free regulatory protein, and the degree of availability (to the regulatory protein) of various competing DNA sites, must all be known for the system at hand. Approaches to various aspects of this problem are under development in our laboratory and elsewhere.

ACKNOWLEDGEMENTS

The research from our laboratory that is reviewed here has been supported by USPHS Research Grants GM-15792 and GM-29158 (to PHvH), and by partial salary support from the Swedish National Science Research Council (to OGB). An earlier version of this chapter was published in *DNA–Ligand Interactions*, W. Guschlbauer and W. Saenger (eds), Plenum, New York (1987), and those parts are reproduced with permission here.

REFERENCES

Anderson, J. E., Ptashne, M. and Harrison, S. C. (1987). Structure of the repressor–operator complex of bacteriophage 434. *Nature*, **326**, 846–852
Berg, O. G. and von Hippel, P. H. (1987). Selection of DNA binding sites by regulatory

proteins. Statistical–mechanical theory and applications to operators and promoters. *J. Mol. Biol.*, **193**, 723–750

Berg, O. G. and von Hippel, P. H. (1988). Selection of DNA binding sites by regulatory proteins. II. The binding specificity of cyclic AMP receptor protein (CRP) to recognition sites. *J. Mol. Biol.*, in press

Berg, O. G., Winter, R. B. and von Hippel, P. H. (1982). How do genome-regulatory proteins locate their DNA target sites? *Trends Biochem. Sci.*, **7**, 52–55

deHaseth, P. L., Lohman, T. M. and Record, M. T., Jr (1977). Nonspecific interaction of *lac* repressor with DNA: an association reaction driven by counterion release. *Biochemistry*, **16**, 4783–4790

deHaseth, P. L. Lohman, T. M., Record, M. T., Jr and Burgess, R. R. (1978). Nonspecific interactions of *Escherichia coli* RNA polymerase with native and denatured DNA: differences in the binding behavior of core and holoenzyme. *Biochemistry*, **17**, 1612–1622

Dickson, R. C., Abelson, J. N., Barnes, W. M. and Reznikoff, W. S. (1975). Genetic regulation. *lac* control region. *Science*, **182**, 27–35

Fairfield, F. R., Newport, J. W., Dolejsi, M. K. and von Hippel, P. H. (1983). On the processivity of DNA replication. *J. Biomol. Struct. Dyn.*, **1**, 715–727

Goeddel, D. V., Yansura, D. G. and Caruthers, M. H. (1978). How *lac* repressor recognizes *lac* operator. *Proc. Natl Acad. Sci. USA*, **75**, 3578–3582

Kopka, M. L., Yoon, C., Goodsell, D., Pjura, P. and Dickerson, R. E. (1985). The molecular origin of DNA-drug specificity in netropsin and distamycin. *Proc. Natl Acad. Sci. USA*, **82**, 1376–1380

Kowalczykowski, S. C., Lonberg, N., Newport, J. W. and von Hippel, P. H. (1981). Interactions of T4 coded gene 32-protein with nucleic acids. I. Characterization of the binding interactions. *J. Mol. Biol.*, **145**, 75–104

Liu-Johnson, H.-N., Gartenberg, M. R. and Crothers, D. M. (1986). The DNA binding domain and bend angle of *E. coli* CAP protein. *Cell*, **47**, 995–1005

Melchior, W. B., Jr and von Hippel, P. H. (1973). Alteration of the relative stability of dA·dT and dG·dC base pairs in DNA. *Proc. Natl Acad. Sci. USA*, **70**, 298–302

Mossing, M. C. and Record, M. T., Jr (1985). Thermodynamic origins of specificity in the *lac* repressor–operator interactions. Adaptability in the recognition of mutant operator sites. *J. Mol. Biol.*, **186**, 295–305

O'Gorman, R. B., Rosenberg, J. M., Kallai, O. B., Dickerson, R. E., Itakura, K., Riggs, A. D. and Matthews, K. S. (1980). Equilibrium binding of inducer to *lac* repressor–operator DNA complex. *J. Biol. Chem.*, **255**, 10107–10114

Ohlendorf, D. H., Anderson, W. F., Fisher, R. G., Takeda, Y. and Matthews, B. W. (1982). The molecular basis of DNA-protein recognition inferred from the structure of *cro* repressor. *Nature*, **298**, 718–723

Ptashne, M., Jeffrey, A., Johnson, A. D., Maurer, R., Meyer, B. J., Pabo, C. O., Roberts, T. M. and Sauer, R. T. (1980). How the λ repressor and cro work. *Cell*, **19**, 1–11

Record, M. T., Jr, deHaseth, P. L. and Lohman, T. M. (1977). Interpretation of monovalent and divalent cation effects on the *lac* repressor–operator interaction. *Biochemistry*, **16**, 4791–4796

Record, M. T., Jr, Lohman, T. M. and deHaseth, P. L. (1976). Ion effects on ligand–nucleic acid interactions. *J. Mol. Biol.*, **107**, 145–158

Revzin, A. and von Hippel, P. H. (1977). Direct measurement of association constants for the binding of *Escherichia coli lac* repressor to non-operator DNA. *Biochemistry*, **16**, 4769–4776

Schneider, T. D., Stormo, G. D., Gold, L. and Ehrenfeucht, A. (1986). Information content of binding sites on nucleotide sequences. *J. Mol. Biol.*, **188**, 415–431

Seeman, N. C. Rosenberg, J. M. and Rich, A. (1976). Sequence-specific recognition of double-helical nucleic acids by proteins. *Proc. Natl Acad. Sci. USA*, **73**, 804–808

Tronrud, D. E., Holden, H. M. and Matthews, B. W. (1987). Structures of two thermolysin-inhibitor complexes that differ by a single hydrogen bond. *Science*, **235**, 571–574

von Hippel, P. H. (1979). On the molecular bases of the specificity of interaction and transcriptional proteins with genome DNA. In Goldberger, R. F. (ed.), *Biological Regulation and Development*, Plenum, New York, 279–347

von Hippel, P. H., Bear, D. G., Winter, R. B. and Berg, O. G. (1982). Molecular aspects of promoter function: An overview. In Chamberlain, M. and Rodriquez, R. (eds), *Promoters: Structure and Function*, Praeger, New York, 3–33

von Hippel, P. H. and Berg, O. G. (1986). On the specificity of DNA–protein interactions. *Proc. Natl Acad. Sci. USA*, **83**, 1608–1612

von Hippel, P. H. and Berg, O. G. (1989). Facilitated target location in biological systems. *J. Biol. Chem.*, **264**, 675–678 (minireview)

von Hippel, P. H. and Fairfield, F. R. (1985). Thermodynamic aspects of the regulation of protein synthesis in bacteria. *Pure Appl. Chem.*, **57**, 45–56

von Hippel, P. H. and McGhee, J. D. (1972). DNA–protein interactions. *Ann. Rev. Biochem.*, **41**, 231–300

von Hippel, P. H., Revzin, A., Gross, C. A. and Wang, A. C. (1974). Non-specific DNA binding of genome regulating proteins as a biological control mechanism. I. The *lac* operon: Equilibrium aspects. *Proc. Natl Acad. Sci. USA*, **71**, 4808–4812

Weintraub, H. and Groudine, M. (1976). Chromosomal subunits in active genes have an altered conformation. *Science*, **193**, 848–856

Winter, R. B., Berg, O. G. and von Hippel, P. H. (1981). Diffusion-driven mechanisms of protein translocation on nucleic acids. III. The *E. coli lac* repressor–operator interaction: Kinetic measurements and conclusions. *Biochemistry*, **20**, 6961–6977

Winter, R. B. and von Hippel, P. H. (1981). Diffusion-driven mechanisms of protein translocation on nucleic acids. II. The *E. coli lac* repressor–operator interaction: equilibrium measurements. *Biochemistry*, **20**, 6948–6960

Woodbury, C. P., Jr, Hagenbuchle, O. and von Hippel, P. H. (1980). DNA site recognition and reduced specificity of the *Eco*RI endonuclease. *J. Biol. Chem.*, **255**, 11534–11546

Woodbury, C. P., Jr and von Hippel, P. H. (1981). Relaxed sequence specificities of *Eco*RI endonuclease and methylase: Mechanisms, possible practical applications and uses in defining protein–nucleic acid recognition mechanisms. In Chirikjian, J. (ed.), *The Restriction Enzymes*, Elsevier, Amsterdam, 181–207

Yarus, M. (1969). Recognition of nucleotide sequences. *Ann. Rev. Biochem.*, **38**, 841–880

2
Structures of protein–nucleic acid complexes in solution by electro-optical analysis

Dietmar Porschke and Jan Antosiewicz

I INTRODUCTION

Although methods for the analysis of biomolecular structures have been developed up to a remarkably high degree of sophistication, and have also been applied with great success, it is still a considerable problem to obtain detailed information on the structure of biological macromolecules in solution. The most powerful method for structure analysis is, of course, X-ray diffraction. However, this method is restricted to the investigation of crystals and structures found in the crystalline state need not be equivalent to those in solution. In principle, a detailed analysis of structures in solution is possible by NMR measurements, but in practice this method is still limited to the analysis of relatively small molecules. Thus, other methods have to be applied, if information on the structure of large macromolecules in solution is required. In this chapter, we give a short description of electro-optical methods, which prove to be particularly useful for the analysis of protein–nucleic acid complexes. We start from a short account of experimental procedures and of theoretical foundations for a quantitative analysis. We then discuss applications, which are of general interest in the domain of protein–nucleic acid interactions. Finally, a short summary is given of results obtained by electro-optical procedures and on the potential of this method for future investigations of protein–nucleic acid complexes.

II ELECTRO-OPTICAL EXPERIMENTS

The experimental procedure used for electro-optical investigations is

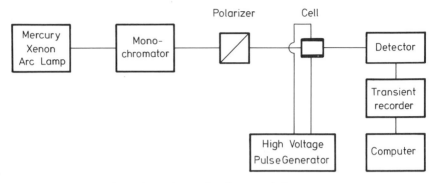

Figure 2.1   Scheme of experimental setup for electro-optical measurements

relatively simple: samples are subjected to electric field pulses, and the response due to field-induced alignment or field-induced reactions is recorded by spectrophotometric techniques (Fredericq and Houssier, 1973). In many cases, the term electro-optics is used in a more restricted sense for the induction of anisotropy via molecular alignment by electric field pulses and its quantitative analysis by linear dichroism or birefringence.

All the data discussed below have been measured by an electro-optical apparatus, which is shown schematically in Figure 2.1. Voltage pulses applied to the electrodes of the cell induce partial alignment of the macromolecules in solution, provided that the macromolecules are associated with a permanent dipole moment or have a preferential polarizability in one direction. If the macromolecules are also anisotropic with respect to their absorbance, the alignment can be followed by measurements of the absorbance of polarized light. The results are usually represented in the form of the 'linear dichroism':

$$\xi = \frac{\Delta A\| - \Delta A_\perp}{A},$$

where $\Delta A\|$ and $\Delta A_\perp$ are the absorbance changes of light polarized parallel and perpendicular to the field vector, respectively; A is the isotropic absorbance. Measurements of the linear dichroism as a function of the electric field strength can then be used to evaluate both electric and optical properties of the molecules under investigation. The electric properties can be dominated by an asymmetric distribution of charges, which is then reflected by a permanent dipole moment. Some molecules, like DNA, show a particularly large polarizability along one axis, which leads to orientation according to an induced dipole mechanism. Evaluation of the stationary degree of orientation as a function of the electric field strength according to the appropriate orientation mechanism provides the limit optical anisotropy corresponding to complete alignment. The limit linear dichroism, which can be evaluated by this procedure, is a direct measure of the orientation of the chromophores with respect to the alignment axis.

An independent source of information are the time constants τ_i of rotational diffusion, which can be determined after pulse termination by recording the transition from the partially aligned state to the random distribution.

$$\xi(t) = \Sigma Q_i e^{t/\tau_i}$$

where Q_i are the amplitudes associated with the individual rotational relaxation processes. The interpretation of these time constants is relatively simple, when the shape of the molecules corresponds to that of simple geometric bodies like spheres, elipsoids or cylinders. In most cases of protein–nucleic acid complexes, however, the shapes are more complex, and thus quantitative evaluation of rotation time constants requires more sophisticated procedures. Since these procedures are hardly known, but prove to be quite useful, we give a short account of the theoretical background below.

III SHORT DESCRIPTION OF HYDRODYNAMIC MODEL CALCULATIONS

Rotational diffusion coefficients reflect molecular dimensions with particularly high sensitivity. For example, the rotational diffusion coefficient of a sphere (cf. Cantor and Schimmel, 1980)

$$D_r = \frac{kT}{6\eta V} = \frac{kT}{8\pi\eta r^3} = \frac{kT}{\zeta_{rot}} \tag{2.1}$$

decreases with the inverse of the third power of the radius r ($\eta \equiv$ viscosity, $kT \equiv$ thermal energy, $V \equiv$ volume of the sphere; $\zeta_{rot} \equiv$ rotational friction coefficient). For comparison, the translational diffusion of a sphere

$$D_t = \frac{kT}{6\pi\eta r} = \frac{kT}{\zeta_{trans}} \tag{2.2}$$

decreases only with the inverse of the radius (ζ_{trans} is the translation friction coefficient). Similar differences are found for the diffusion coefficients of other objects like ellipsoids or rods. Thus, rotational diffusion coefficients are most sensitive indicators of molecular dimensions. In the past, applications to analysis of molecular structures have been limited by difficulties in the interpretation of data obtained for objects, which cannot be modelled by any of the simple geometric bodies. However, this limitation is not serious any more, thanks to the development of appropriate hydrodynamic models and the availability of sufficient computing capacity. The approach used for our analysis of protein–nucleic acid complexes was initiated in

1927 by Oseen and was later developed by Burgers, Kirkwood *et al.* and some other groups. The subject has been reviewed in 1981 by Garcia de la Torre and Bloomfield, who also contributed to the development of the theory with respect to the analysis of biopolymers. The theory is based on the hydrodynamics of simple spherical beads. Complex shapes are generated by arrangements of such beads and the main issue of the theory is then the calculation of hydrodynamic interactions between these beads. The frictional force experienced by each bead of a moving bead assembly is described by Stokes' law in the form

$$\vec{F}_i = -6\pi\eta r_i(\vec{u}_i - \vec{v}_i) \tag{2.3}$$

where $\vec{u}_i$ is the velocity of the ith bead and $\vec{v}_i$ is the velocity that the solvent would have at the centre of the ith bead, if this bead were absent. In the limit case of a molecule modelled by one bead only, $\vec{v}_i = 0$, provided that the solvent does not move in the absence of the molecule. However, in the presence of the other beads, $\vec{v}_i \neq 0$. This effect on the friction experienced by the ith bead, which is transmitted from the other beads by the solvent, is called 'hydrodynamic interaction'. The first quantitative treatment of these interactions was given by Oseen (1927). When the forces on each bead are calculated, the translational friction coefficient is determined by the total force, corresponding to the sum of the individual ones, at unit velocity of the whole assembly.

A corresponding procedure is used to describe the rotation of a bead assembly at constant angular velocity $\vec{\omega}$. In this case, the linear velocity of the ith bead is given by

$$\vec{u}_i = \vec{\omega} \times \vec{a}_i \tag{2.4}$$

where $\vec{a}_i$ is the position vector of the ith bead in the coordinate system, which is defined to describe the bead assembly. The force is then determined by equation (2.3), and thus the torque $\vec{T}_i$ experienced by the ith bead is given by

$$\vec{T}_i = \vec{a}_i \times \vec{F}_i \tag{2.5}$$

Summation of all terms $\vec{T}_i$ provides the total torque, which is equivalent to the appropriate rotational friction coefficient at unit angular velocity.

According to this procedure, a bead does not contribute to the rotational friction coefficient, if it is located on the rotation axis. However, the rotational friction coefficient of a bead with radius r rotating around a central axis is given by

$$\zeta_{rot} = 8\pi\eta r^3 = 6\eta V \tag{2.6}$$

where V is the volume of the sphere. For this reason, Garcia de la Torre and Rodes proposed to add a correction given by equation (2.6) to the principal rotational friction coefficients calculated for the bead assembly, where V is the volume of the assembly.

Since the torques for each bead involve their position vectors (cf. equation 2.5), the calculated rotational friction coefficients depend on the choice of the origin of the coordinate system used to describe the model. The coefficients have physical significance only when calculated with respect to a special point called the centre of hydrodynamic resistance. A procedure for the correct assignment of this point has been given by Brenner and by Harvey together with Garcia de la Torre.

At present, the theoretical basis has been developed for calculations of translational and rotational friction coefficients on rigid bead assemblies with non-overlapping beads of various sizes or with overlapping beads of constant bead radius. The approach has been checked extensively for simple bead models, where analytical or exact numerical results are available. In general, the agreement is good, or even very good, for translational friction coefficients, and good, or at least satisfactory, for rotational friction coefficients. Thus, rotational diffusion coefficients, which are derived from appropriate friction coefficients, can be calculated for a given model with at least a satisfactory absolute accuracy. The models described below have been calibrated by using rotation time constants, which have been determined experimentally for the major components. Owing to this procedure, our conclusions depend on the relative accuracy, which is clearly higher than the absolute accuracy.

We have to add that the rotational relaxation time constants, which are determined by electro-optical experiments, have been calculated from the (rotational) diffusion coefficients using analytical expressions derived by Wegener *et al.* (1979). For particles without symmetry, there are five separate rotational time constants. Existence of symmetry elements leads to a reduction of this number. Rotation of a sphere, for example, is associated with a single time constant. The amplitudes of individual rotational relaxation modes depend on various factors, including the optical anisotropy, the rotational diffusion coefficients and – for the case of electro-optical data – on the electrical anisotropy. The assignment of all these factors can be rather difficult. In the case of protein–nucleic acid complexes, the assignment is simplified by the fact that the optical anisotropy is clearly dominated by the contribution of the nucleic acid component. A corresponding statement holds for the electrical anisotropy, although an exact treatment requires a careful analysis of a potential contribution by the protein component. More detailed discussions of all the procedures used for the present simulations are given by Garcia de la Torre and Bloomfield (1981), and also by Antosiewicz and Porschke (1988).

IV   REPRESSOR PROTEINS WITH LARGE PERMANENT DIPOLE MOMENTS OF ABOUT 1000 DEBYE UNITS

Two repressor proteins have been analysed recently by electro-optical measurements. In both cases, the electrical and optical anisotropies were sufficiently high for accurate quantitative evaluations (Porschke, 1987; Porschke *et al.*, 1988). The linear dichroism, measured as a function of the electrical field strength, is clearly not compatible with an alignment due to induced dipole moments, but can be represented only by orientation due to permanent dipoles. The existence of permanent dipole moments is also shown by the kinetics of the dichroism rise under electric field pulses. Thus, there can be little doubt as to the correct assignment of permanent dipoles. We emphasize this conclusion, because the magnitude of the permanent dipole moments appears to be unbelievably high. In the case of lac repressor the dipole moment is about 4×10^{-27} C m (corresponding to 1200 Debye units), and for Tet repressor a dipole moment of 3.5×10^{-27} C m (corresponding to 1050 Debye units) has been found.

It should be mentioned that dipole moments of biological macromolecules have been discussed with great caution, mainly because the magnitude of the dipoles could not be determined with sufficient accuracy. Furthermore, the nature of the observed dipole moments has not been clear in all cases. The method used in the present case provides the magnitude of the dipole moments without assumptions, and the conclusion on the nature of the dipole has been secured by additional measurements, which will not be discussed in the present context because of space limitations.

A permanent dipole moment, of course, implies an asymmetric molecular construction, which has been found by X-ray analysis for several specific DNA binding proteins. Although the X-ray structures of lac- and Tet repressors are not yet available, we may conclude from our present results that their structures are asymmetric as well. This is not so surprising for Tet repressor, but was not expected for lac repressor, which is usually considered to be a symmetric tetramer. However, the nature of the asymmetry in the case of lac repressor remains to be established.

Another consequence of the high permanent dipole moment is expected for the interactions of these protein molecules. It is well known that much smaller dipole moments, like that of water, for example, have a considerable impact on their intermolecular interactions. Obviously, the dipole moment of repressors should be useful for their interactions with DNA molecules. First of all, the DNA double helix is highly polarizable, which should lead to considerable dipole-induced dipole interactions. In addition, the positive end of the protein dipole is of course constructed for strong interactions with the negative phosphates of DNA. These interactions are non-specific, but should also be useful to direct the protein into the proper orientation for recognition of its specific DNA sequence.

Electro-optical measurements also provide information on the shape of the protein molecules from their rotational diffusion time constants. According to these data, both lac and Tet repressor are elongated molecules (Porschke, 1987; Porschke *et al.*, 1988). The high sensitivity of rotational time constants with respect to molecular dimensions can be used to test for changes of structure due to inducer binding. Addition of the inducers isopropyl-β-D-thiogalactopyranoside to lac repressor and tetracycline to Tet repressor did not result in any detectable change of the rotation time constant, which rules out any major change of the 'external' structure. Changes in the 'internal' structure due to inducer binding, however, are demonstrated by variations of the limit dichroism.

V   DNA DOUBLE HELICES MODELLED BY BEADS

The electrical properties of DNA double helices are very complex, owing to their polyelectrolyte nature, and thus will not be discussed in the present context. As should be expected, double helices have a very high polarizability, which can be used to align them along their axis by electric field pulses and to study their rotational diffusion after pulse termination. The discussion below will be restricted to the quantitative interpretation of experimental rotation time constants by appropriate hydrodynamic models. Since we want to extend the interpretation to protein–DNA complexes, standard models of DNA helices like simple rods are not useful, and the more complex representation by bead assemblies introduced above has to be applied. Thus, the first task is construction of a bead assembly, which represents the hydrodynamics of double helices. Again we do not discuss details, and simply mention that we start from a string of overlapping beads, with external dimensions corresponding to those found for double helices in crystals. In solution, double helices are of course hydrated and surrounded by a cloud of counterions. Thus, the hydrodynamic dimensions are expected to be larger than the crystal dimensions. This expectation is verified by our simulations, which show that the initial dimensions of our string of beads have to be blown up in order to fit the observed rotation time constants. Because the hydration and the ion environment are likely to be similar in all directions, we increased the radius of the beads and the length of the string simultaneously by the same increments. The experimental dependence of time constants on buffer composition and concentration is reflected by some variation of the hydrodynamic dimensions (Antosiewicz and Porschke, 1988; Porschke *et al.*, 1988). Since our bead models do not consider any flexibility, applications are limited to chain lengths below the persistence length.

VI TET REPRESSOR–OPERATOR COMPLEX

Rotation time constants have been determined from electric dichroism experiments for the Tet repressor, specific operator DNA fragments and repressor–operator complexes (Porschke *et al.*, 1988). Thus, bead models could be constructed independently for the protein and the DNA components. For simulation of the rotational diffusion of complexes, we simply have to align protein and DNA beads into close contact, with the centre of the protein at the centre of the palindromic DNA sequence. Some overlap between the components is allowed – up to a maximal value of 3 Å. Each of the two operators, which are arranged in tandem on the DNA fragments, is combined with one protein dimer. Since the resulting complexes are not symmetric, more than a single rotational time constant is computed. However, one of these time constants is associated with almost all of the dichroism amplitude, and the other relaxation components do not contribute more than about 1 per cent. The exact distribution of the dichroism amplitudes might be influenced by a potential contribution of the protein dipole to the orientation mechanism, which is dominated by the large DNA polarizability. Model calculations with several orientations of a protein dipole showed that the slow process prevails in all cases of practical interest. Another complication may result from the relative position of the two protein dimers around the DNA, which is not exactly known. However, alignment of the proteins at opposite sides of the DNA, rather than at the same side, leads to an increase of the rotation time constant by not more than about 5 ns.

As already mentioned above, the dichroism decay is mainly determined by a simple exponential. Nevertheless, we have considered the influence of the other components by the fitting of the calculated decay curves by single exponentials $\bar{\tau}$, using a least-squares fitting procedure corresponding to that applied for the analysis of the experimental data. Given in this form, theoretical and experimental results are completely equivalent. The calculated $\bar{\tau}$ values for complexes between Tet repressor and straight operator DNA helices are somewhat larger than the corresponding experimental values, which have been obtained for several operator DNA fragments. For one of these fragments, it is known from gel retardation experiments that the repressor does not bind quantitatively, whereas binding is complete for the other fragments. Complete agreement of theoretical and experimental results for the cases with complete binding can be obtained by some bending of the DNA. When we assume smooth bending, we get excellent agreement of simulated and experimental time constants for a radius of curvature 500 Å. As illustrated in Figure 2.2, this corresponds to a relatively small degree of bending. Because the accuracy of both simulated and experimental time constants is limited to approximately ±10 ns, we cannot completely rule out that DNA remains straight. The

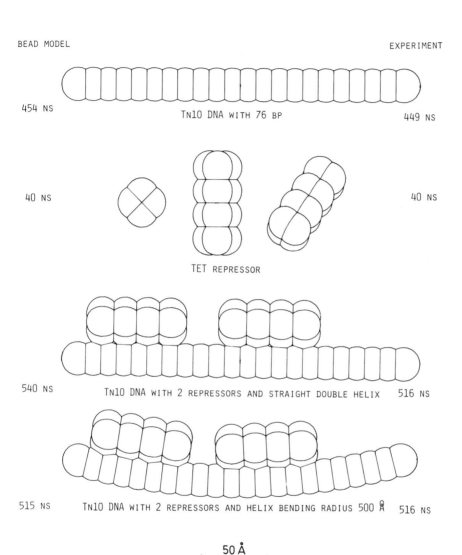

BEAD MODEL EXPERIMENT

454 NS

TN10 DNA WITH 76 BP

449 NS

40 NS 40 NS

TET REPRESSOR

540 NS TN10 DNA WITH 2 REPRESSORS AND STRAIGHT DOUBLE HELIX 516 NS

515 NS TN10 DNA WITH 2 REPRESSORS AND HELIX BENDING RADIUS 500 Å 516 NS

50 Å

Figure 2.2 Bead models of Tn10 DNA with 76 bp, the Tet repressor protein in three different views, the complex formed from two repressors with straight Tn10 DNA and the same complex after bending of the DNA at a bending radius of 500 Å. The dichroism decay time constants for the bead model and the corresponding experimental values are on the left and right side, respectively (Porschke *et al.*, 1988). Since the rotation time constant of the complex is mainly determined by the long dimension of the DNA, we cannot rule out the possibility that the repressor dimers are located on opposite sides of the double helix (cf. text)

other limit, corresponding to the maximum degree of bending consistent with the experimental data, is a radius of curvature ~400 Å.

We may conclude that binding of Tet repressor to its operator does not induce a major change of the long-range DNA structure. Bending of the double helix – if existent at all – is limited to a relatively small degree, and thus the two repressors, which are located relatively close to each other under saturation of the tandem operator, can hardly come into contact with each other.

VII   COMPLEXES FORMED BY cAMP RECEPTOR WITH PROMOTOR DNA FRAGMENTS

The cyclic AMP receptor protein controls the expression of several genes of *E. coli* by a special activation mechanism (de Crombrugghe *et al.*, 1984). Various observations indicate that binding of this protein to specific DNA induces bending of the double helix (Kolb *et al.*, 1983; Wu and Crothers, 1984). The most direct experimental evidence for DNA bending comes from dichroism decay measurements, which show a strong reduction of rotational diffusion time constants under specific conditions (Porschke *et al.*, 1984). A quantitative interpretation of these time constants, however, requires application of detailed hydrodynamic theory and at present can be given only on the basis of bead models.

In the case of the cAMP receptor protein, rotation time constants are not available, but the detailed structure of this protein is known from X-ray analysis (McKay and Steitz, 1981). Thus, the assembly of beads for hydrodynamic simulation of the protein has been based on the crystal structure. For technical reasons, the radius of beads used for the DNA and the protein component have to be equivalent. The bead assembly constructed for the protein under these boundary conditions is shown in three orthogonal views in Figure 2.3.

The DNA has been attached to this bead model with the centre of the DNA palindrom at the centre of the two helix–turn–helix motifs of the protein dimer. The attachment at the centres of the binding sites is very similar to that designed by Weber and Steitz (1984). Because of the asymmetry of the complexes, the simulation results in five rotational relaxation times. In most cases, one of the rotational modes is dominant; in some cases, more rotational relaxation modes are associated with detectable amplitudes, but nevertheless have not been detected in the experimental decay curves because of very similar magnitudes of their time constants. Thus, the comparison of theoretical and experimental time constants is based again on single exponentials obtained by least-squares fitting (cf. previous section).

Before we start a detailed comparison of experimental and theoretical results, we should first present some general conclusions, which are evident

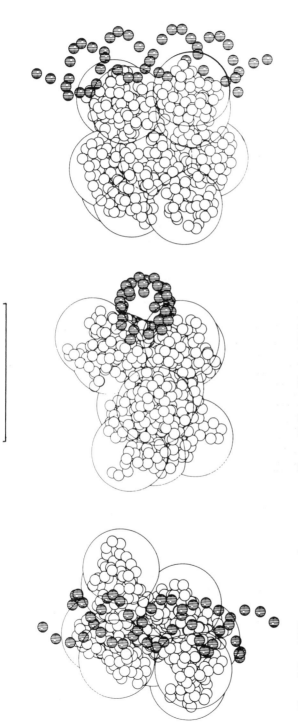

Figure 2.3   cAMP receptor protein structure according to McKay and Steitz (1981) in three orthogonal views with a bead model consisting of 13 beads superimposed. The small hatched beads show the DNA helix bound to the protein according to the model of Weber and Steitz (1984)

50 Å

from the experimental results without numerical simulations. As shown by the data compiled in Table 2.1, binding of the protein to DNA fragments in the absence of cAMP does not lead to much change of the rotation time constant with respect to that of free DNA. Obviously, the rotation time constants of the complexes are dominated by the DNA component because of its particularly large external dimensions. The rather small changes observed upon protein binding indicate that these large DNA dimensions are not much affected in the complexes. Addition of the small inducer cAMP, however, leads to very large reductions of the rotation time constants for several complexes, which can be explained only by a considerable reduction of the external DNA dimensions. This reduction is dependent not only on the presence of cAMP but also on the ionic strength: the changes are particularly large at higher salt concentrations. According to these results, we will not be able to assign one unique structure for the cAMP receptor–promotor complex, but apparently have a broad spectrum of possible structures and a strong dependence of the average conformation existing in solution on the environmental conditions.

Table 2.1   Complexes of cAMP receptor protein with various DNA fragments: rotation relaxation times in μs at 20 °C from dichroism decay

bp	Binding site	Buffer	DNA	DNA + CRP	DNA + CRP + cAMP
98	non-specific	T	0.88	0.91	0.88
98	non-specific	T10	0.81	0.84	0.81
62	specific 1	T	0.33	0.43	0.25
80	specific 2	T	0.57	0.56	0.56
203	specific 1 + 2	T	4.5	4.7	2.4
80	specific 2	T10	0.49	0.52	0.13
203	specific 1 + 2	T10	3.8	3.7	1.48

Buffer T: 5 mM Tris pH 8, 0.1 mM DTE, 0.1 mM EDTA.
Buffer T10: T + 10 mM NaCl.

From Porschke *et al.* (1984).

Interpretation of the experimental data beyond these qualitative conclusions requires bead model simulations. We will discuss a series of simulations (Antosiewicz and Porschke, 1988) for the complex formed with an 80 bp fragment, which contains a single specific binding site, in the high-salt buffer T10 (cf. Table 2.1). The rotation time constant calculated for a complex with straight DNA (560 ns) is slightly higher than the corresponding experimental value observed in the absence of cAMP. This result indicates that the dimensions of the DNA are reduced already by binding of the protein in the absence of inducer. The most simple procedure for simulation of this reduction is smooth bending of the double helix. As shown in Figure 2.4, smooth bending to a radius of curvature of 600 Å reduces the rotation time constant to 547 ns. A further decrease of

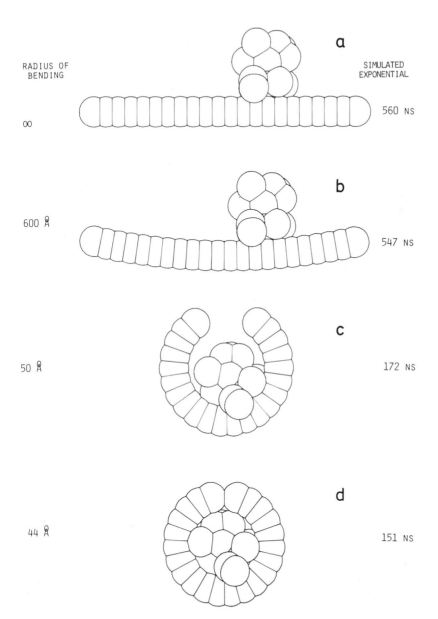

RADIUS OF
BENDING

SIMULATED
EXPONENTIAL

∞

560 NS

600 Å

547 NS

50 Å

172 NS

44 Å

151 NS

a

b

c

d

Figure 2.4   Bead model of cyclic AMP receptor associated with straight 80 bp DNA (a). The same complex at different degrees of smooth bending of the DNA are shown in (b) (600 Å), (c) (50 Å) and (d) (44 Å), together with the simulated dichroism decay time constant. The experimental value is 130 ns (from Antosiewicz and Porschke, 1988)

the radius of curvature to approximately 500 Å leads to a fit of the experimental time constant (520 ns). Much higher degrees of bending are required to simulate the rotational time constant observed in the presence of cAMP. As shown in Figure 2.4, the DNA has to be wrapped around the protein to a very close contact and at a radius of curvature 44 Å, in order to come reasonably close to the experimental time constant of 130 ns. In this case, bending of the DNA approaches the maximal value compatible with the protein structure.

Simulation of the experimental data observed for other DNA fragments and/or for other conditions did not reveal the same extreme degree of bending, but considerable bending of the DNA has been deduced for most specific complexes. A special case is encountered for the fragment with 203 bp, because it contains two specific binding sites for the cAMP receptor. We include a discussion of our simulations for this case (Antosiewicz and Porschke, 1988), because this complex is expected to be relevant for the situation *in vivo*. The distance between the centres of the two sites is 71.5 bp, which is close to 7 helical turns of B-DNA. Thus, the two protein dimers on the DNA palindroms are located almost 'in phase' to the same side of the double helix and may come into close contact upon bending of the DNA. From independent investigations, it is known that the cAMP receptor protein has a tendency to aggregate upon binding to DNA. Furthermore, we may use the well-known fact that 2×2 protein subunits preferentially associate to a tetrahedral geometry. This geometry has the useful feature that a binding site for a DNA double helix, presented by one dimer on one side of the tetrahedron, is given for the second dimer on the other side in a proper orientation, which fits smoothly into the same double helix, when the DNA is bent around, and thus passes the second site in the opposite direction. As indicated already above, this arrangement requires a separation of the binding sites on the DNA by an integer number of helix turns.

Apparently, all these requirements are fulfilled in the case of the complex formed from two cAMP receptor dimers and the 203 bp fragment. The model resulting from these considerations is shown in Figure 2.5, and its rotation time constant is in excellent agreement with the experimental result. However, the complex shown in Figure 2.5 also raises some problems: for example, the bending of the double helix is not symmetric around the centres of the binding sites. Other models may be constructed with a more detailed consideration of local stereochemical constraints. In any case, we may conclude that the overall degree of DNA bending should be similar to that shown in Figure 2.5.

VIII CONCLUSIONS

It is notoriously difficult to get detailed information on macromolecular

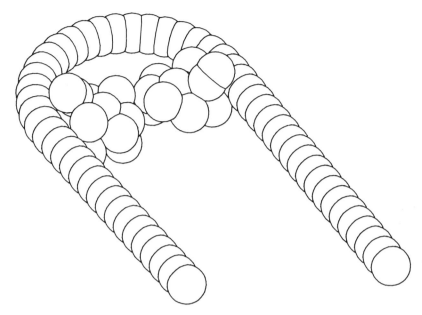

Figure 2.5   Bead model of the complex formed from a DNA fragment of 203 bp with two specific sites and two cAMP receptor protein dimers. The simulated time constant for the dichroism decay is 1.43 μs and the corresponding experimental value is 1.48 μs (from Antosiewicz and Porschke, 1988)

structures in solution. None of the methods available provides all the required information by simple experimental procedures that are applicable in general. Although NMR methods are usually considered to be most promising in this respect, applications are still restricted to molecules of relatively low molecular weight. Under these conditions, information on macromolecular structures in solution has to be collected by various methods. In this chapter, we want to demonstrate the potential of electro-optical procedures. As shown by the examples discussed above, the electro-optical approach can be particularly useful for the analysis of protein–nucleic acid complexes. However, the full potential of this method can be exploited only by appropriate hydrodynamic model calculations.

It is well known that rotational diffusion coefficients are very strongly dependent on external dimensions. Experimental data can be interpreted relatively easily for molecules with simple symmetric shapes like spheres or rods, but problems arise for molecules with a more complex shape. Owing to the development of detailed hydrodynamic theories, these problems can be solved. In this respect, protein–nucleic acid complexes are particularly susceptible to a detailed analysis, because the optical properties of the complexes in the UV are dominated by the contribution of the nucleic acid component, and the optical anisotropy of complexes is defined by the folding pathway of the DNA double helix, for example. A similar argument, though not quite as rigorous, holds for the electric anisotropy. Under these conditions, models can be calculated in considerable detail.

The problem of DNA folding by proteins is of general interest, and appears to be particularly important for the regulation of gene activity. The present analysis of experimental data obtained for two helix–turn–helix proteins shows a remarkable difference of their complexes with specific DNA. The long-range structure of operator DNA remains almost unaffected upon complexation with its repressor, whereas the promotor DNA is strongly bent by complex formation with its gene activator. Obviously, these conclusions may not hold in general. Before general conclusions are justified, investigations of the structure of more protein–nucleic acid complexes in solution are required. Electro-optical experiments, together with hydrodynamic simulations, prove to be particularly useful for this purpose.

REFERENCES

Antosiewicz, J. and Porschke, D. (1988). Turn of promotor DNA by cAMP receptor protein characterized by bead model simulation of rotational diffusion. *J. Biomol. Struct. Dynamics*, **5**, 819–837

Cantor, C. R. and Schimmel, P. R. (1980). *Biophysical Chemistry*, Freeman and Co., San Francisco

de Crombrugghe, B. Busby, S. and Buc, H. (1984). Cyclic AMP receptor protein: role in transcription activation. *Science*, **224**, 831–838

Fredericq, E. and Houssier, C. (1973). *Electric Dichroism and Electric Birefringence*, Clarendon, Oxford

Garcia de la Torre, J. and Bloomfield, V. A. (1981). Hydrodynamic properties of complex, rigid, biological macromolecules: theory and applications. *Quart. Rev. Biophys.*, **14**, 81–139

Kolb, A., Spassky, A., Chapon, C., Blazy, B. and Buc, H. (1983). On the different binding affinities of CRP at the lac, gal and malT promotor regions. *Nucleic Acids Res.*, **11**, 7833–7852

McKay, D. B. and Steitz, T. A. (1981). Structure of catabolite gene activator protein at 2.9 Å resolution suggests binding to left-handed B-DNA. *Nature*, **290**, 744–749

Oseen, C. W. (1927). Hydrodynamik in *Mathematik und ihre Anwendungen in Monographien und Lehrbüchern*, Hilb. E., ed., Akad. Verlagsges., Leipzig

Porschke, D., Hillen, W. and Takahashi, M. (1984). The change of DNA structure by specific binding of the cAMP receptor protein from rotation diffusion and dichroism measurements. *EMBO J.*, **3**, 2873–2878

Porschke, D. (1987). Electric, optical and hydrodynamic parameters of lac repressor from measurements of the electric dichroism. High permanent dipole moment associated with the protein. *Biophys. Chem.*, **28**, 137–147

Porschke, D., Tovar, K. and Antosiewicz, J. (1988). Structure of Tet repressor and Tet repressor–operator complexes in solution from electrooptical measurements and hydrodynamic simulations. *Biochemistry*, **27**, 4674–4679

Weber, I. T. and Steitz, T. A. (1984). Model of specific complex between catabolite gene activator protein and B-DNA suggested by electrostatic complementarity. *Proc. Natl Acad. Sci. USA*, **81**, 3973–3977

Wegener, W. A., Dowben, R. M. and Koester, V. J. (1979). Time-dependent birefringence, linear dichroism, and optical rotation resulting from rigid-body rotational diffusion. *J. Chem. Phys.*, **70**, 622–632

Wu, H. M. and Crothers, D. M. (1984). The locus of sequence-directed and protein-induced DNA bending. *Nature*, **308**, 509–513

3
NMR studies of protein–DNA recognition. The interaction of *lac* repressor headpiece with operator DNA

Robert Kaptein, Rolf Boelens and Rolf M. J. N. Lamerichs

INTRODUCTION

Repressors are proteins that regulate transcription by binding to DNA control regions (the so-called operators). The classic system for regulation of gene expression is the *lac* operon of *E. coli*. Many ideas on negative control of transcription, and, indeed, the now familiar concept of an operon, were originally derived from a study of the *E. coli* lactose genes (Jacob and Monod, 1961). *Lac* repressor was also the first repressor that was isolated (Gilbert and Müller-Hill, 1966). Over the years, it has been the subject of numerous biochemical and genetic studies (for reviews see Bourgeois and Pfahl, 1976; Miller and Reznikoff, 1978; Miller, 1979; 1984). Initially, the *lac* repressor–*lac* operator system served as the principal model system for studies on specific DNA recognition by proteins. When it appeared that *lac* repressor refused to crystallize in a form suitable for X-ray crystallography, it lost its primacy and attention was turned to other proteins involved in gene regulation such as CAP and the phage λ repressors, cI and cro. The crystal structures of these proteins were solved in the early 1980s (McKay and Steitz, 1981; Pabo and Lewis, 1982; Anderson *et al.*, 1981). This crystallographic work and its implications for protein–DNA recognition have been reviewed by Pabo and Sauer (1984). Later, the structure of the *trp* repressor was solved (Schevitz *et al.*, 1985) and, at low resolution, that of a complex of 434 repressor and its cognate operator (Anderson *et al.*, 1987).

From these studies, combined with model building, it became clear that prokaryotic repressors use a helix–turn–helix structural motif for recognition of operators (Pabo and Sauer, 1984). In particular, amino acid side chains of the second α-helix of this domain (the recognition helix) interact with nucleic acid bases in the major groove of DNA. Evidence for this comes also from genetic studies, such as the elegant helix swapping experiments by Wharton and Ptashne (1985), in which the binding specificity of 434 repressor was changed to that of P22 repressor. Interactions of amino acid residues from the recognition helices with specific nucleic acid bases could be demonstrated for CAP (Ebright *et al.*, 1984), *lac* repressor (Ebright, 1985), λ and cro repressors (Hochschild and Ptashne, 1986; Hochschild *et al.*, 1986), and 434 repressor (Wharton and Ptashne, 1987).

Meanwhile, *lac* repressor is making a comeback. The three-dimensional structure of its N-terminal DNA binding domain (or 'headpiece') has been determined by two-dimensional NMR methods (Kaptein *et al.*, 1985), and extensive genetic characterizations are going on in the laboratories of J. H. Miller and B. Müller-Hill. In this article, we review the NMR work of our laboratory on the structure of *lac* headpiece and its complexes with *lac* operator fragments (for a review of the early NMR work, see Kaptein *et al.*, 1983). Although *lac* repressor follows the current paradigm, in that it also has a helix–turn–helix DNA binding domain, the NMR results have shown that it uses this domain in a different way. Its recognition helix binds in the major groove in an orientation that is opposite to that of all other repressors with known structures (Boelens *et al.*, 1987a), a result that has recently been confirmed in genetic experiments (Müller-Hill, private communication; Lehming *et al.*, 1987). Thus, it appears that repressors fall into two classes, depending on the orientation of the recognition helix with respect to the dyad axis of the repressor–operator complex.

BIOMOLECULAR STRUCTURES FROM NMR

NMR methodology

It is now possible to determine the three-dimensional structure of small biomolecules (up to MW 20000) in solution using NMR. Powerful two-dimensional (2D) NMR techniques have been developed (for a review, see Ernst *et al.*, 1987), which make possible the ^1H resonance assignments of both small proteins and oligonucleotides (Wüthrich, 1986). Progress has also been made in developing methods for structure determination on the basis of constraints derived from NMR, such as distance geometry (Havel *et al.*, 1979, 1983; Braun and Go, 1985) and restrained molecular dynamics (Van Gunsteren *et al.*, 1983; Kaptein *et al.*, 1985; Clore *et al.*, 1985).

The primary source of information on which NMR structures of biomolecules are based is the nuclear Overhauser effect (NOE). In 1D experiments, it is defined as the change in NMR intensity of one nucleus, which occurs when another is magnetically saturated. In 2D NMR, NOEs are manifested as off-diagonal cross-peak intensities in a 2D NOE spectrum, recorded in a three-pulse experiment (Jeener *et al.*, 1979)

$$90° - t_1 - 90° - t_m - 90° - t_2 \text{ (acq.)}$$

In this pulse sequence, 90° stands for a 90° radiofrequency pulse, t_1 and t_2 are the variable times which after double Fourier transformation yield the ω_1 and the ω_2 frequency domains of a 2D spectrum. Acquisition of the free induction decays takes place during t_2; t_m is a fixed time, which allows exchange of magnetization between nuclei. The origin of the NOE effect is dipolar cross-relaxation, which depends on fluctuations in the orientation and length of the vectors connecting pairs of nuclei. In a rigid molecule, these vectors have fixed lengths and reorient by the tumbling of the molecule as a whole. In that case, cross-relaxation rates are proportional to r^{-6} and therefore have a very strong distance dependence. As an example, a 2D NOE spectrum of *lac* repressor headpiece in D_2O is shown in Figure 3.1. All off-diagonal intensity in this spectrum corresponds to short distances between non-exchangeable protons. Using suitable calibration distances (e.g. between neighbouring protons in an aromatic ring), proton–proton NOEs can then be translated into distances, which in turn can be used as constraints in structure determination procedures.

For real biomolecules in solution, the assumption of rigidity, of course, is not correct, and local motions are indeed a major source of errors in distance measurements from NOEs. Errors may also arise from indirect magnetization transfer, or 'spin diffusion', which may occur when the mixing time t_m in a 2D NOE experiment is not very short. Thus, proton–proton distances up to 4 or 5 Å can be determined from 2D NOE spectroscopy with an accuracy of 10 per cent at best. Apart from distance information, J-coupling may provide constraints on dihedral angles. The coupling constants can be conveniently measured from spacings in the cross-peak fine structure of 2D NMR spectra. They can be translated into dihedral angles, using relations such as that proposed by Karplus (1959). However, only for small (2–4 Hz) or large (9–10 Hz) J-couplings are these relations unambiguous. For intermediate values, uncertainties may arise, first, because the Karplus curves are multivalued in the sense that several dihedral angles may belong to a certain J-coupling, and, second, because they may be the result of motional averaging.

Thus, the type of information on which NMR structures are based is short-range in nature: short (<5 Å) distances from NOEs and dihedral angles from J-couplings. Since these short distance constraints come with errors, one has to worry about the propagation of errors in structures determined from a few hundred of these constraints. In globular molecules, such as proteins, where a linear chain falls back on itself possibly a few times, this appears not to be a serious problem. In fact, overall

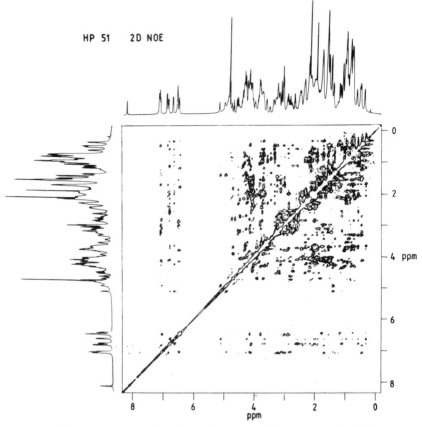

HP 51   2D NOE

Figure 3.1   2D NOE spectrum of headpiece 51 represented as a contour plot. The spectrum was recorded at 500 MHz of a 5 mM solution of headpiece in D_2O. The mixing time t_m was 100 ms. Off-diagonal cross-peaks indicate short distances (<4 Å) between non-exchangeable protons. 1D spectra are shown along the ω_1 and ω_2 axes

properties of small proteins such as the radius of gyration can be obtained from NMR with surprising accuracy (Wagner *et al.*, 1987). For a 'linear' molecule, such as a DNA fragment, the situation is less favourable. Here one would expect that local conformations can well be determined from NMR data, but long-range properties less accurately.

Structure determination from NMR data

A set of distance and dihedral angle constraints having been obtained from NMR (typically several hundreds for a small protein or oligonucleotide), the next question is how to derive the three-dimensional structure. While a single generally accepted method for this does not exist, and various authors have used different procedures, we believe that the protocol shown in Scheme 1 may serve for this purpose. It has been applied to the structure determination of *lac* repressor headpiece. The first two steps, consisting of ^{1}H resonance assignment and determination of distance and dihedral angle constraints, are common to all procedures. Steps 3 and 4 are suitable to address questions such as, How unique are the structures obtained? How well do they satisfy the experimentally derived constraints? and How reasonable are they from the point of view of energetics? The question of uniqueness is an important one, especially since the information obtainable from NMR is often less extensive and less accurate than that from X-ray crystallography. Therefore, it is necessary to calculate families of structures, with the optimal sampling of the conformation space that is consistent with the constraints.

Scheme 1   Protocol for biomolecular structure determination from NMR

---

1. Assign ^{1}H resonances.
2. Determine proton–proton distance constraints and dihedral angle constraints from NOEs and J-couplings, respectively.
3. Calculate family of structures using geometric constraints only (experimental constraints plus covalent structure), using, for instance, distance geometry (DG) and distance bounds driven dynamics (DDD).
4. Refine these structures using geometric constraints and potential energy functions, for instance, with restrained energy minimization (REM) and restrained molecular dynamics (RMD).

The metric matrix distance geometry (DG) algorithm (Blumenthal, 1970; Havel *et al.*, 1979, 1983; Havel and Wüthrich, 1985) calculates structures from geometric constraints in the form of upper and lower bounds on atom–atom distances. It is the only method that does not require some starting conformation, and is therefore free from operator bias. It contains a random step in choosing a set of distances between upper and lower bounds, and therefore different structures can be obtained. Yet it was found that DG structures tend to cluster around extended conformations and do not sample the available conformation space completely randomly (Havel and Wüthrich, 1985; Scheek and Kaptein, 1989). Adding a so-called distance-bounds-driven dynamics (DDD) step after the DG calculations improves the sampling properties dramatically (Scheek and Kaptein, 1989). This amounts to a simplified molecular dynamics calculation, in which the molecule is allowed to move under an artificial force-field that contains only a distance constraint error function and no energetic terms. Starting in a DG conformation, the molecule is given some kinetic energy and a DDD run then results in a large spread in conformations still consistent with the bounds. Application of the DG + DDD procedure to *lac* headpiece has been described by Scheek and Kaptein (1989).

Next, one has the option of refining the structures, including a potential energy function. A procedure has been worked out to combine experimental constraints in the form of a pseudo-potential with the force-fields used in conventional molecular mechanics or molecular dynamics calculations (McCammon and Harvey, 1987). This restrained molecular dynamics (RMD) procedure often results in improved structures, which have both a lower internal energy and fewer constraint violations. It should be borne in mind, however, that these structures now depend on the quality of the force-field and may have certain artifacts, especially when solvent is not included in the calculations. A more detailed discussion of the methodology of structure determinations based on NMR data is outside the scope of this chapter, but can be found, for instance, in recent reviews by Wüthrich (1986), Kaptein *et al*. (1988) and Scheek and Kaptein (1989).

STRUCTURE OF *LAC* REPRESSOR HEADPIECE

Lac repressor, a tetrameric protein of molecular weight 154 000, is too large for high-resolution NMR studies. However, each subunit has a separate DNA-binding domain (headpiece) that can be cleaved off by proteolytic enzymes (Geisler and Weber, 1977). The amino acid sequence of the N-terminal region is shown in Figure 3.2. Depending on the proteolytic enzyme used, headpieces can be prepared containing 51, 56 or 59 amino acid residues (HP 51, HP 56 or HP 59). These headpieces retain their original three-dimensional structure and their ability to recognize the *lac* operator specifically (Ogata and Gilbert, 1979). The trypsin-resistant core is involved in the subunit interaction and contains the inducer binding site. The sequence of the natural *lac* operator reveals an approximate two-fold symmetry (Gilbert and Maxam, 1973) and in agreement with that, two subunits of *lac* repressor suffice to recognize *lac* operator (Kania and Brown, 1976). Therefore, *lac* repressor should bind with two headpieces to each half of the operator.

[1]H-resonance assignments and secondary structure

The structure elucidation of *lac* headpiece started with the assignments of its ^{1}H resonances. Prior to the 2D NMR work, the tyrosine resonances had been assigned using methods such as selective nitration (Ribeiro *et al*., 1981) and comparison with genetically altered headpieces (Arndt *et al*., 1981). Application of the photo-CIDNP method (Kaptein, 1982) to *lac* headpiece showed that His 29 and tyrosines 7, 12 and 17 are surface-accessible residues, while Tyr 47 is a buried one (Buck *et al*., 1980).

Using a combination of 2D NMR experiments, it is now possible to make assignments for virtually all proton resonances in a small biomolecule (Wüthrich, 1986). This so-called sequential assignment procedure makes use of the two main classes of 2D NMR experiments: one in

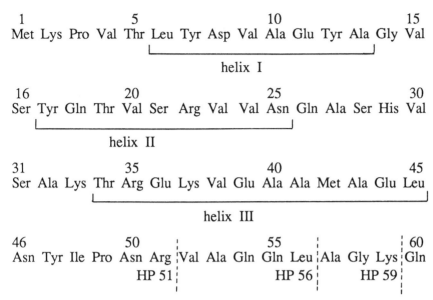

| 1 | | | | 5 | | | | | 10 | | | | | 15 |
| Met | Lys | Pro | Val | Thr | Leu | Tyr | Asp | Val | Ala | Glu | Tyr | Ala | Gly | Val |

helix I

| 16 | | | | 20 | | | | | 25 | | | | | 30 |
| Ser | Tyr | Gln | Thr | Val | Ser | Arg | Val | Val | Asn | Gln | Ala | Ser | His | Val |

helix II

| 31 | | | | 35 | | | | | 40 | | | | | 45 |
| Ser | Ala | Lys | Thr | Arg | Glu | Lys | Val | Glu | Ala | Ala | Met | Ala | Glu | Leu |

helix III

46				50					55					60
Asn	Tyr	Ile	Pro	Asn	Arg	Val	Ala	Gln	Gln	Leu	Ala	Gly	Lys	Gln
				HP 51					HP 56			HP 59		

Figure 3.2   Amino acid sequence of the N-terminal part of *lac* repressor (Beyreuther *et al.*, 1973). The α-helical regions as determined by NMR are indicated. Cleavage sites are shown of the enzymes clostripain, chymotrypsin and trypsin, yielding headpiece fragments of 51, 56 and 59 amino acid residues, respectively

which cross-peaks arise only between protons connected through J-coupling networks, with the correlated spectroscopy or COSY experiments as the prime example, and the other one with cross-peaks between protons that are spatially in close proximity, the 2D NOE or NOESY experiment. COSY and related spectra, therefore, represent the covalent structure of the molecule, while 2D NOE spectra provide a map of short proton–proton distances. The analysis usually starts with a search for cross-peak patterns belonging to the spin-systems of each type of amino acid. These are then connected through cross-peaks in a 2D NOE spectrum, which represent short distances along the backbone, $d_{\alpha N}$, $d_{\beta N}$, d_{NN} (distances between the C_α, C_β and amide protons of one residue and the amide proton of the next residue in the chain, respectively). This procedure is described extensively in the monograph by Wüthrich (1986). Applied to *lac* headpiece (HP 51), it yielded ^{1}H resonance assignments for all backbone C_α and amide protons, with the exception of those of Ile 48, and for the great majority of the side chain protons (Zuiderweg *et al.*, 1983a; Zuiderweg *et al.*, 1985a).

With the ^{1}H assignments, known cross-peaks in the 2D NOE spectra such as that of Figure 3.1 can be identified. It is convenient to distinguish short- and medium-range NOEs (those between amino acid residues not more than four apart in the sequence) and long-range NOEs. The former provide information about the secondary structure of the protein. This is because the various secondary structure elements (α-helix, β-sheet and turns) have their own characteristic short distances. For headpiece stretches of strong amide–amide NOEs and NOEs between C_α-protons and amides of the third and fourth residue further in the chain, $d_{\alpha N}(i, i+3)$ and $d_{\alpha N}(i, i+4)$, gave strong evidence for the presence of three α-helical regions, helix I for residues 6–13, helix II for 17–25 and helix III for 34–45 (Zuiderweg *et al.*, 1983b). The first two helices correspond to the famous helix–turn–helix structural motif that had been predicted for *lac* repressor on the basis of sequence homology with other repressors (Ohlendorf *et al.*, 1982; Sauer *et al.*, 1982). The position of the helices is indicated in Figure 3.2.

Tertiary structure

The next step is to determine the overall folding of the protein. For HP 51, the structure determination is based on a set of 169 NOEs (Zuiderweg *et al.*, 1985a). Care was taken to eliminate the effect of spin-diffusion by examining 2D NOE spectra taken at relatively short mixing times (50 ms and 100 ms). The NOEs were converted to upper-bound distance constraints of 3.5 Å corresponding to the distances, $d_{\alpha N}$ and $d_{\alpha N}(i, i+3)$, in regular α-helices that were used for calibration. The initial structure determination of HP 51 then consisted of a model building step followed by refinement using the restrained molecular dynamics approach (Kaptein *et al.*, 1985). First, a molecular model was built on the basis of a number of key NOEs, using the α-helices as building blocks. This model was then refined with the RMD procedure, which in fact was first applied to the present problem (Kaptein *et al.*, 1985; Zuiderweg *et al.*, 1985b).

The result of this exercise was a structure that satisfied almost all distance and dihedral angle constraints, and at the same time had a reasonably low energy (De Vlieg *et al.*, 1986). Figure 3.3 shows a stereo picture of a snapshot taken from the RMD run. The helix–turn–helix region consisting of the helices I and II of the headpiece can be clearly seen with the third helix packing against the first two forming a hydrophobic core. The RMD run also indicated that the three-helical core of the protein is rather rigid, whereas the N-terminal and C-terminal region and also the loop between helices I and III showed higher mobility.

To address the question of uniqueness a series of DG and DDD calculations was performed (Kaptein *et al.*, 1988; Scheek and Kaptein, 1989). The variation among a family of 10 structures can be expressed as an average r.m.s. difference of the C_α-atom coordinates, which was 2.0 Å for the C_αs of residues 4–47 (3.0 Å for all C_α-atoms). The conformation of the three N-terminal and four C-terminal residues is not very well defined by the constraints, which is most probably due to a high mobility. Refinement

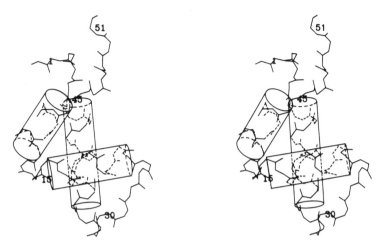

Figure 3.3   Stereo diagram of the backbone conformation of headpiece 51. Cylinders represent the three α-helices. The structure was determined from a set of 169 proton–proton distance constraints from NOEs, using a restrained molecular dynamics procedure (De Vlieg *et al.*, 1986)

of these structures with restrained MD resulted in some convergence for the helical core of the protein; the average r.m.s. difference for the C_{α}s (4–47) became 1.7 Å, while for all C_{α}-atoms it remained at 3.0 Å.

Thus, the NMR studies have shown that the DNA binding domain of *lac* repressor basically has a three-helical structure, resembling the helices 2, 3 and 4 of λ cI repressor (Pabo and Lewis, 1982). The first two helices have a relative orientation, which is surprisingly similar to that of λ and cro repressors and CAP (Steitz *et al.*, 1982).

LAC OPERATOR FRAGMENTS

Lac operator of *E. coli* is defined genetically as the control region in the *lac* operon, where operator constitutive mutants occur. The region protected by *lac* repressor is 20–25 bp long, with a pseudo-dyad axis going through GC 11 (Gilbert and Maxam, 1973). It was found by Sadler *et al.* (1983) and Simons *et al.* (1984) that symmetrical *lac* operators lacking the central GC base-pair bind *lac* repressor up to an order of magnitude stronger than the native one. The sequences of the operators and the fragments discussed here are shown in Figure 3.4. Most of the NMR work has been done with a 14 bp operator fragment comprising the left half of *lac* operator. This was a fortunate choice, since it turned out to be the stronger binding half, which occurs also in the symmetrical operator (cf. Figure 3.4b).

Initially, NMR work on the 14 bp *lac* operator fragment focused on the imino protons of the G and T bases (Zuiderweg *et al.*, 1981). These protons resonate in a separate region of the spectrum (12–15 p.p.m.), and are

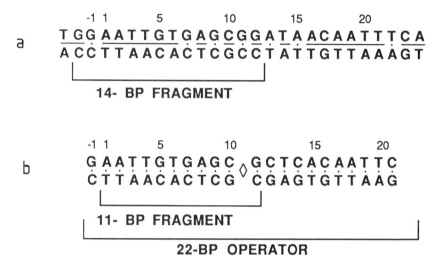

Figure 3.4 Sequences of native *lac* operator (a) (Gilbert and Maxam, 1973) and 'ideal' symmetric *lac* operator (b) (Sadler *et al.*, 1983; Simons *et al.*, 1984). Synthetic operator fragments of 11, 14 and 22 bp used in the NMR studies are indicated

relatively easily accessible for NMR study. Because they are involved in Watson–Crick base pairing and are situated in the interior of the helix, they are good structural probes for DNA melting and ligand binding. Assignments for the imino protons were obtained from melting studies and comparison with 7 bp subfragments. They were later confirmed by NOE experiments (Kaptein *et al.*, 1983). Lu *et al.* (1983) measured the imino proton exchange rates of a number of operator fragments. They noted an exceptionally high exchange rate for the central T in GTG sequences (T6 in the *lac* operator). Whether this is related to the recognition of these sequences by proteins is not known.

With the development of 2D NMR methods, it became possible to obtain virtually complete ^{1}H assignments of double-stranded oligonucleotides of 20 base pairs or more (Scheek *et al.*, 1983a, 1984; Hare *et al.*, 1983). The following procedure can be used. First, a COSY spectrum is recorded, showing cross-peaks between protons of the same ribose unit and between cytosine H5 and H6 protons (the only J-coupled protons in the bases). Then, from 2D NOE spectra, sequential assignments are made by searching for cross-peaks connecting neighbouring nucleotides. Here, the assumption is made initially that the DNA is a right-handed helix. However, this assumption is confirmed (or rejected) at a later stage, so the assignments do not depend on the assumed handedness. In right-handed DNA, short distances prevail from a base H6 or H8 protons (for pyrimidines and purines, respectively) to ribose H1', H2' and H2'' protons, both within a nucleotide unit and to ribose protons on the neighbouring nucleotide at the 5' side (not at the 3' side). Pathways of connected proton–proton distances, which are short enough for cross-relaxation, run through the whole strand of DNA. One example of such cross-relaxation pathways involving ribose H1', H2' and H2" and the base H6/H8 protons is shown in Figure 3.5 for the 14 bp *lac* operator fragment (Scheek *et al.*, 1985). The lines in region *a* (sometimes called the fingerprint region) connect intranucleotide H1'–H6/H8 cross-peaks with internucleotide ones. Since several different cross-relaxation paths exist also involving H3' and H4' and H5 (cytosine) and 5-methyl protons (thymine), many checks for consistency can be made, which usually lead to unambiguous assignments. For the 14 bp operator fragment, all non-exchangeable protons were assigned in this way, except for some H5' and H5" protons, which often display little shift dispersion.

When 2D NOE spectra of an oligonucleotide are recorded in H_2O, assignments can be obtained for the imino and amino protons as well (Boelens *et al.*, 1985). The amino protons are close to non-exchangeable protons (H5 for cytosine, H2 for adenine). Therefore, networks of NOE cross-peaks can be traced connecting them to these protons, which can first be assigned from D_2O spectra. Cytosine amino protons occur as sharp doublets, but those of guanine and adenine are, under normal conditions, collapsed to (broad) singlets, owing to exchange processes, and appear to be less useful.

The 11 bp and 22 bp operator fragments that were recently studied (Boelens *et al.*, 1988; Lamerichs, Boelens and Kaptein, unpublished results) could be assigned in a similar way. The NOE cross-peak intensities for all these fragments showed clearly that their conformations are in the B-DNA family. More detailed structure determinations have not yet been attempted. With the great majority of protons assigned for both *lac* headpiece and *lac* operator fragments, and their basic structures known, the stage is set for studies of their mutual interaction.

LAC HEADPIECE – OPERATOR COMPLEXES

NMR studies

The complex of HP 51 with the 14 bp operator fragment has a molecular weight of *ca.* 14000, which makes it one of the largest systems to have been studied in detail by high-resolution NMR. The initial work dealt with the imino protons of the DNA and aromatic resonances of the protein, which are relatively easily accessible (Scheek *et al.*, 1983b). It was shown that HP 51 binds to the operator fragment in a specific way, forming a complex that is in fast exchange with the constituents. However, analysis of the binding curves showed that, apart from the specific interactions, non-specific complexes are formed as well.

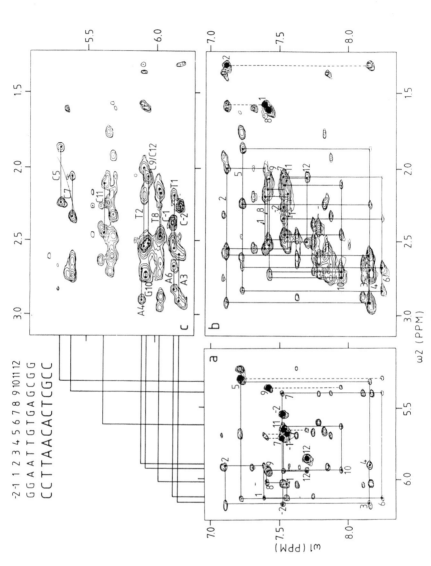

Figure 3.5   Networks of NOE cross-peaks in the 2D NOE spectrum of the bottom strand in the 14 bp *lac* operator fragment. Lines connect intra- and internucleotide cross-peaks between H6/H8 and H1′ (in region a), between H6/H8 and H2′ and H6/H8-H2″ (in region b), and between H′ and H2′ and H2″ (in region c). Dashed lines connect cross-peaks involving cytosine H5 and thymine 5-methyl protons

The imino protons shifted up to 0.2 p.p.m., but remained all visible in the spectrum, indicating that the double helix remains intact upon complex formation and undergoes relatively small conformational changes. Similar shift changes were observed for a 51 bp DNA fragment containing the full *lac* operator (Buck *et al.*, 1983). A stoichiometry of two headpieces per operator was deduced (Nick *et al.*, 1982; Buck *et al.*, 1983), in agreement with fluorescence and circular dichroism studies (Culard *et al.*, 1982).

The aromatic headpiece residues show characteristic chemical shift changes upon operator binding mainly for Tyr 7, Tyr 17 and His 29 (Nick *et al.*, 1982; Scheek *et al.*, 1983b). His 29 becomes protonated when the titration is carried out at pH 6.8. This residue increases its pK_a by 0.5 unit when bound to the operator, indicating that it is involved in an ionic interaction, probably with a backbone phosphate on the DNA (Scheek *et al.*, 1983b).

From photo-CIDNP experiments, similar conclusions could be drawn. Complexes of HP 51 with poly [d(AT)] (Buck *et al.*, 1980) and with the 14 bp operator fragment (S. Stob, R. M. Scheek, R. Boelens and R. Kaptein, unpublished results) showed reduced photo-CIDNP response for His 29, Tyr 7 and Tyr 17, indicating that access to these residues is blocked by DNA. By contrast, the CIDNP effect for Tyr 12 remained unchanged, so that this residue must remain accessible in the complex. Nick *et al.* (1982) monitored *lac* repressor operator interaction using ^{19}F NMR spectroscopy. The tyrosines in the repressor were substituted by 3-fluorotyrosines, which had been assigned by genetic means (Jarema *et al.*, 1981). Examination of the ^{19}F chemical shift changes that occur upon DNA binding lead qualitatively to the same conclusions as the results from ^1H NMR: very small shift changes are involved in the non-specific interaction with DNA, while operator binding induces significant shift changes for tyrosines 7 and 17.

Figure 3.6 shows the 500 MHz spectra of free headpiece 56, the 11 bp operator and a 1:1 complex in D_2O. From the figure, it is clear that there will be a number of 'windows' in a 2D spectrum, where no overlap of protein and DNA resonances exists. These areas of the 2D spectrum lend themselves to start the assignment of the resonances of headpiece and operator in the complex. The full 2D NOE spectrum of this complex is shown in Figure 3.7. It is clear that the interpretation of spectra of this level of complexity is a formidable task, and the analysis has taken several years (Boelens *et al.*, 1987a, 1987b).

Figure 3.8 shows one of the most readily accessible regions of a 2D NOE spectrum of the complex of HP 56 with the 14 bp operator. It contains a window, where only intra-DNA cross-peaks occur (H6/H8–H1′ and cytosine H5–H6). These cross-peaks provided a start for the assignment of the DNA resonances, as is shown in Figure 3.8 for one strand by the lines connecting intra- and internucleotide cross-peaks. In this way, assignments

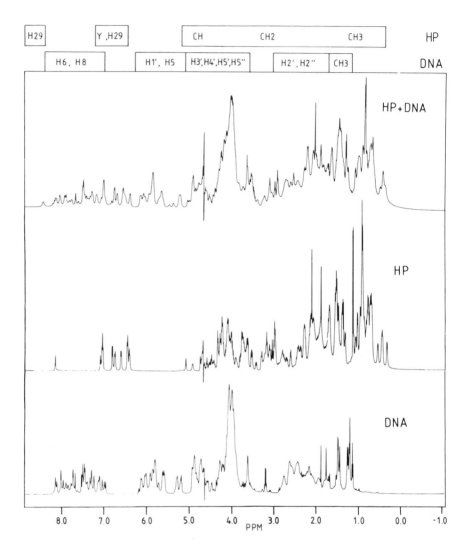

Figure 3.6   500 MHz ¹H NMR spectra of 11 bp *lac* operator fragment (bottom), headpiece 56 (middle) and the HP 56–11 bp operator complex (top). Concentrations are 4 mM in both protein and DNA in D₂O. For further conditions see Boelens *et al*. (1987b). Regions where types of protein and DNA protons occur in the spectrum are indicated

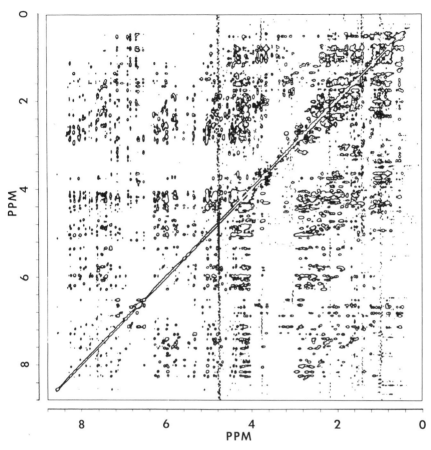

Figure 3.7   500 MHz ^{1}H 2D NOE spectrum of HP 56–11 bp operator complex taken with a mixing time of 300 ms

were obtained for all non-exchangeable protons of the DNA in the complex, except for some of the H5′ and H5″ protons. The general pattern of intra-DNA NOEs is still that of a B-DNA type conformation. Also, most of the ^{1}H resonance positions show small shifts upon complex formation, with a maximum of 0.2 p.p.m. for the H8 proton of G5 and the H1′ proton of G7. These results are also consistent with the idea that small adjustments of the DNA conformation occur. These conformational changes, however, cannot yet be specified, but are likely to involve bending or unwinding or a combination of both.

Similarly, a large number of ^{1}H assignments have been made for the protein part of the HP 56–14 bp operator complex. For this, a combination of 2D NOE spectra and homonuclear Hartman–Hahn (HOHAHA) spectra was used. In the latter experiment, cross-peaks are observed between J-coupled spins, like in COSY, but now connecting more than one pair of protons in the side chains of amino acid residues. The protons of many internal residues such as Leu 6, Val 9, Leu 45 and Tyr 47 have characteristic chemical shifts (distinct from random coil), which change very little upon complex formation. Furthermore *ca.* 80 per cent of the long range NOEs could be identified for headpiece in the complex. This shows that the basic three-helical structure of headpiece is conserved when it binds to the operator. Shifts occur only for residues in the DNA binding site, which may be due to the presence of the DNA (His 29) or to a repositioning of the side chains, as probably occurs for Tyr 7 and Tyr 17.

Now what can we say about headpiece–operator interaction? Using the assignments of the 14 bp operator and of HP 56 in the complex, it was possible to detect NOEs between protein and DNA. Some of these can be seen as cross-peaks in the 2D NOE spectrum of Figure 3.8. For instance, extending the horizontal line at 7.96 p.p.m. (of the H8 proton of G10) to low field, one finds a cross-peak at 6.53 p.p.m., which can belong only to the 3, 5 protons of Tyr 7. The analysis so far has yielded 24 protein–DNA NOEs for the HP 56–14 bp complex. Similarly, in the 2D NOE spectrum of the complex, with the 11 bp operator, 21 NOEs could be observed between protein and DNA (Boelens *et al.*, 1988). In the case of the 22 bp symmetrical operator (cf. Figure 3.4), two headpieces were bound. Here 16 interprotein–DNA NOEs could be identified, which, except in one case, were a subset of those already observed in the other complexes. These results are collected in Table 3.1.

It should be mentioned that the NOEs listed in the table were observed in 2D NOE spectra taken at rather long mixing times of 100 ms and 250 ms. This was done in order to facilitate the assignments. Some of the NOE cross-peaks may be the result of indirect magnitization transfer (spin diffusion). For instance, it is unlikely that Leu 6 lies in the DNA binding site and protein–DNA NOEs involving this residue are weak and probably contain contributions of spin diffusion via Tyr 7. For the same reason, the distances corresponding to these NOEs may be up to 6 Å. In Table 3.1, the distinction is made between NOEs that are unambiguous because they involve protons with unique resonance positions and those that are probable. The latter ones occur in crowded regions, where overlap of resonances may occur. They were assigned on the basis of a pattern recognition procedure, which involves the following reasoning. Suppose a cross-section of a headpiece proton shows NOEs to a set of other protons of the same amino acid residue. Then, if a cross-section of a DNA proton shows cross-peaks at the same frequencies, and at least one of these can be uniquely assigned to a headpiece proton, we consider the assignment of the other cross-peaks in the set to also be extremely likely. The case of His 29 may serve as an example. In Figure 3.8, a horizontal line at the C2 proton frequency of His 29 (8.52 p.p.m.) shows a set of four cross-peaks in the DNA ribose region which are in a crowded region of the spectrum. Now, a

Table 3.1   NOEs observed between headpiece 56 and operator fragments
of 11, 14 and 22 base pairs

Protein		DNA		11 bp	14 bp	22 bp
unambiguous[a]						
Tyr 7	H3, 5	-G10	H8	+	+	+
Tyr 7	H2, 6	-G10	H8	+		+
Tyr 7	H3, 5	-G10	H1'		+	
Tyr 7	H3, 5	-G10	H3'	+	+	+
Tyr 7	H3, 5	-C9	H5	+	+	+
Tyr 7	H3, 5	-C9	H6	+	+	+
Leu 6	C_8H_3	-C9	H5	+	+	
Thr 5	$C_\alpha H$	-G10	H3'	+		
Tyr 17	H3, 5 + H2, 6	-C9	H5	+	+	+
Tyr 17	H3, 5 + H2, 6	-C9	H6		+	
Tyr 17	H3, 5 + H2, 6	-T8	H6	+	+	+
His 29	H2	-A2	H8		+	+
His 29	H2	-T3	CH_3	+	+	+
probable[a]						
Thr 5	$C_\gamma H_3$	-G10	H8	+	+	+
Thr 5	$C_\gamma H_3$	-G10	H3'	+	+	
Leu 6	C_8H_3	-C9	H5		+	
Leu 6	C_8H_3	-C9	H6	+	+	
Leu 6	C_8H_3	-C9	H3'		+	
Leu 6	C_8H_3	-T8	H6		+	
Tyr 17	H3, 5 + H2, 6	-T8	CH_3	+	+	
Ser 21	$C_\alpha H$	-T8	CH_3	+	+	
Ser 21	$C_\beta H$	-T8	CH_3	+	+	
His 29	H2	-A2	H3'	+	+	+
His 29	H2	-A2	H4'	+	+	+
His 29	H2	-A2	H5'[b]	+	+	+
His 29	H2	-A2	H5"[b]	+	+	+
His 29	H2	-A2	H1'			+
His 29	H4	-T3	CH_3	+		+

[a] The unambiguous NOEs were assigned at unique resonance frequencies, while the probable
NOEs were from resonances that could overlap with resonances of other protons (see text for
further discussion).
[b] H5' and H5" protons were only pair-wise assigned.

similar set of cross-peaks is observed at the line of H8 of adenine 2, and,
moreover, a very weak cross-peak is observed at the crossing of this line
and at that of His 29, both of which have unique resonance positions (not
shown). Hence, the weak NOE between H8 of A2 and the His 29 C2
proton is listed in the upper part of Table 3.1, although it undoubtedly is
the result of spin diffusion. The other NOEs of His 29, with the ribose
protons of A2, are stronger and represent shorter distances, but occur in a
region of the spectrum with much overlap, and are therefore indicated as
probable.

Inspection of Table 3.1 shows that the NOEs observed for the three
complexes are substantially the same. It is significant that, with the
exception of NOEs involving Leu 6, this is also true for the 2 : 1 complex of
HP 56 and the 22 bp operator, where protein–protein contacts may be

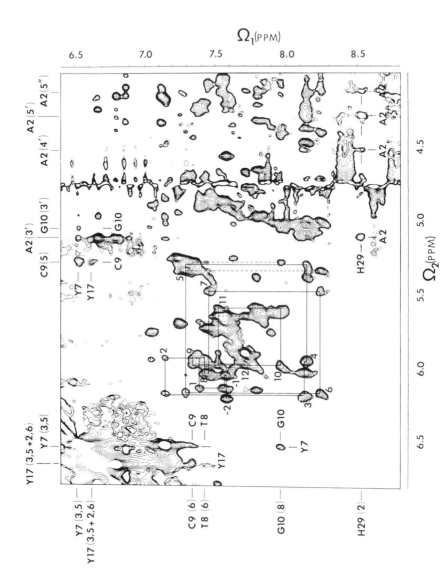

Figure 3.8 Part of the 500 MHz 2D NOE spectrum of the HP 56–14 bp operator complex. Sequential NOEs of the same strand as that of Figure 3.5 are indicated by the connecting lines. The protein–DNA NOE cross-peaks present in this part of the spectrum are also indicated

present. In this case, the headpieces must bind in a very similar way, as in the complexes with the smaller operator fragments.

Structural model for the headpiece–operator complex

The NMR results showed that both headpiece and operator change their conformations only slightly upon binding. It is therefore reasonable to start modelling the complex using the NOE information of Table 3.1, while keeping the operator in a standard B-DNA conformation and headpiece in its NMR derived structure (Zuiderweg *et al.*, 1985b). As mentioned above, the 2D NOE spectra were taken under conditions of limited spin diffusion, so that for the weak NOEs upper bound distance constraints were set at the rather long distance of 6 Å. For the strong NOEs, the upper bound was taken as 4 Å. Where necessary, pseudo-atom corrections were applied. Some of the amino acid side chains in the DNA binding region of headpiece were allowed to change their conformation. Models for the HP 56–14 bp operator were then built, first on a graphics display system (Boelens *et al.*, 1987a, 1987b) and later using the so-called ellipsoid algorithm (Billeter *et al.*, 1987; Boelens *et al.*, 1988). Energy minimization of this model ensured that it had reasonable non-bonded interactions. In this way, a model of the complex was obtained (shown in Figure 3.9) in which all protein–DNA NOE constraints could be simultaneously satisfied. This is important, because it virtually excludes the possibility that some of the observed NOEs are due to non-specific complexes. Although these are certainly formed at the high concentrations (5 mM) of the NMR experiments, they do not lead to observable NOEs (inconsistent with the specific complex), most likely because their lifetime is not long enough to allow buildup of NOE intensity.

The most surprising feature of the model is that the orientation of the second or 'recognition' helix in the major groove of DNA with respect to the dyad axis at GC11 is opposite that found in all other models of repressor–operator interaction, either from direct X-ray observation as for 434 repressor (Anderson *et al.*, 1987) or from models built for CAP and λ and cro repressors (Pabo and Sauer, 1984). Indeed, it is also opposite to orientations predicted for *lac* repressor on the basis of the analogy of models for CAP (Weber *et al.*, 1982) and cro repressor (Matthews *et al.*, 1982). In these models, the first helix would be away from the dyad axis, while in the complex shown in Figure 3.9 it is close to it. The model accounts for the phosphate ethylation interference experiments of Gilbert and Maxam (cited in Barkley and Bourgeois, 1978) and also for a functional contact between Gln 18 and GC 7 as found by Ebright (1985) from a genetic 'loss of contact' study involving mutants of both *lac* repressor and operator. It can also account for tyrosine fluorescence quenching data (Culard *et al.*, 1982) and for the CIDNP results (Buck *et*

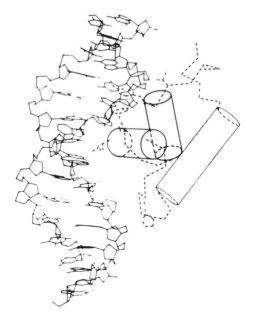

Figure 3.9 Complex of HP 51 with the 14 bp *lac* operator fragment obtained with the ellipsoid algorithm (Boelens *et al.*, 1988). The DNA is in the standard B conformation. The headpiece conformation is that of Zuiderweg *et al.* (1985b), except that some of the side chains in the DNA contact region were allowed to move

al., 1980). One of the strongest protein–DNA NOEs is that between the (overlapping) ring protons of Tyr 17 and the 5-methyl group of T8. It suggests that this is a functional contact, since it is known that the T8 methyl group is essential for repressor binding (Caruthers, 1980). It should be noted that the other protein–DNA NOEs presented in Table 3.1 do not necessarily reflect those residues that are most important for DNA recognition. NMR observation is biased towards some aromatic residues, methyl groups and short side chains that are most easily recognized in the NMR spectrum. Unfortunately, residues like Gln 18 and Arg 22, which must be important in base pair recognition, have long side chains that are difficult to detect.

An important question is whether the whole *lac* repressor binds to the operator with its headpieces in the same orientation as we have found for isolated headpieces. Recently, genetic experiments carried out in the group of B. Müller-Hill have shown that this is actually the case.

Lehming *et al.* (1987) constructed a *lac* repressor mutant with the first two amino acids of the recognition helix replaced by those of *gal* repressor (Tyr 17→Val, Gln 18→Ala). This mutant repressor had high affinity for the *gal* operator, which differs from *lac* operator at positions 7 and 9. Although this already gives some clue as to the orientation of the lac   cro

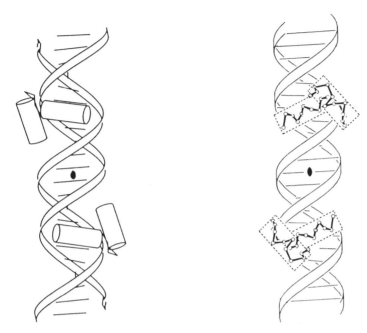

Figure 3.10   Helix–turn–helix domain in repressor–operator complexes for *lac* repressor as determined from NMR (Boelens *et al*. 1988) and as proposed for cro (taken from Pabo and Sauer, 1984). The orientation of the recognition helix is approximately opposite in both models. Note that in the *lac* repressor–operator model this helix is almost perpendicular to the DNA helix axis, while in the proposed cro model it lies more parallel to the direction of the major groove

recognition helix, a more definitive result was their finding of a repressor mutant with Arg 22 replaced by Asn, which now had specificity for a *lac* operator with GC5 replaced by TA (Müller-Hill, private communication). This provides support for the Arg 22–GC5 contact in the native system, which we predicted on the basis of our NMR results (Boelens *et al*., 1987a). It also fixes unambiguously the orientation of the recognition helix as the opposite of that of cro and λ repressors.

CONCLUSIONS

The results of the NMR studies on *lac* headpiece–operator complexes can be summarized as follows.

– The basic three-helical structure of *lac* headpiece does not change upon operator binding, apart from some adjustments of side chains and possibly of the conformation of the loop between helix II and helix III. Similarly, the operator remains in a B-DNA-like conformation. Small changes in conformation are seen, which cannot yet be specified, but may involve bending or unwinding or a combination of both.

– Headpiece binds with its recognition helix in the major groove of DNA. However, contrary to models proposed for the *lac* repressor–operator system (Matthews *et al.*, 1982; Weber *et al.*, 1982), its orientation is the opposite of that deduced for the cro and λ repressors and seen in the crystal for 434 repressor (Pabo and Sauer, 1984; Anderson *et al.*, 1987). This has recently been confirmed for the intact *lac* repressor in genetic experiments involving mutants of both repressor and operator (Lehming *et al.*, 1987; Müller-Hill, private communication). These different modes of binding of the helix–turn–helix motif on the operators are shown in Figure 3.10. It appears, therefore, that there are two classes of helix–turn–helix proteins, which can be generically designated as *lac* and cro. From homology arguments, and also on the basis of the genetic experiments by Lehming *et al.* (1987), it appears that at least the *gal* and *deo* repressors belong to the *lac* class. The repressors with known crystal structures, λ cI, cro, trp, 434 (and the related P 22) and also CAP, fall in the cro class.

– From the model for the *lac* headpiece–operator complex as derived from the NOE constraints, it can be inferred that some specific interactions used in operator recognition. Thus, GC5 probably interacts with Arg 22 with bidentate hydrogen bonds to the guanine base. Support for this interaction comes from recent genetic experiments by Müller-Hill, as discussed above. GC7 can form hydrogen bonds simultaneously with Ser 21 and Gln 18 (evidence for the interaction with Gln 18 has been obtained by Ebright, 1985). The methyl group of T8 is probably engaged in a hydrophobic contact with the aromatic ring of Tyr 17. This can be concluded from the strong NOE of this methyl group with the Tyr 17 ring protons. The OH-groups of Tyr 17 and Tyr 7 may be involved in hydrogen bonding to the base pairs GC 9 and CG 10.

A number of non-specific contacts can also be deduced from the model. His 29 shows numerous NOEs to protons at AT2 and TA3 (cf. Table 3.1). Furthermore, its pK_a is increased by 0.5 unit upon complex formation, indicating an ionic contact near the phosphate between base pairs 2 and 3. Other candidates for non-specific interaction with the sugar–phosphate backbone are Thr 5 and Asn 25.

The amino acid residues mentioned are all known to be necessary for repressor function, since their replacement results in I^- phenotypes (Miller, 1979, 1984). Similarly, substitution of the base pairs, for which we suggest here interactions with the protein, leads to O^c (operator constitutive) mutants (Gilbert *et al.*, 1976).

Further work in our laboratory is aimed at specifying these interactions more precisely. In particular, the observation of exchangeable protons present in hydrogen bonds between protein and DNA, although difficult, would be extremely useful. Nevertheless, it is clear that the NMR studies have already significantly contributed to our understanding of protein–DNA recognition.

REFERENCES

Anderson, J. E., Ptashne, M. and Harrison, S. C. (1987). *Nature (London)*, **326**, 846–849

Anderson, W., Ohlendorf, D., Takeda, Y. and Matthews, B. (1981). *Nature (London)*, **290**, 754–758

Arndt, K. T., Boschelli, F., Lu, P. and Miller, J. H. (1981). *Biochemistry*, **20**, 6109–6118

Barkley, M. D. and Bourgeois, S. (1978). In *The Operon*, 2nd edn, Miller, J. H. and Reznikoff, W. S. (eds), Cold Spring Harbor Press, New York, 177–220

Beyreuther, K., Adler, K., Geisler, N. and Klemm, A. (1973). *Proc. Natl Acad. Sci. USA*, **70**, 3576–3580

Billeter, M., Havel, T. F. and Kuntz, I. D. (1987). *J. Comp. Chem.*, **8**, 132–141

Blumenthal, L. M. (1970). *Theory and Applications of Distance Geometry*, Chelsea, New York

Boelens, R., Scheek, R. M., Dijkstra, K. and Kaptein, R. (1985). *J. Magn. Reson.*, **62**, 378–386

Boelens, R., Scheek, R. M., van Boom, J. H. and Kaptein, R. (1987a). *J. Mol. Biol.*, **193**, 213–216

Boelens, R., Scheek, R. M., Lamerichs, R. M. J. N., de Vlieg, J., van Boom, J. H. and Kaptein, R. (1987b). In *DNA–Ligand Interactions*, Guschlbauer, W. and Saenger, W. (eds), Plenum, New York, 191–215

Boelens, R., Lamerichs, R. M. J. N., Rullmann, J. A. C., van Boom, J. H. and Kaptein, R. (1988). *Prot. Seq. Data Anal.*, **1**, 487–498

Bourgeois, S. and Pfahl, M. (1976). *Adv. Protein Chem.*, **30**, 1–99

Braun, W. and Go, N. (1985). *J. Mol. Biol.*, **186**, 611–626

Buck, F., Rüterjans, H. and Beyreuther, K. (1978). *FEBS Lett.*, **96**, 335–338

Buck, F., Rüterjans, H., Kaptein, R. and Beyreuther, K. (1980). *Proc. Natl Acad. Sci. USA*, **77**, 5145–5148

Buck, F., Hahn, K. D., Zemann, W., Rüterjans, H., Sadler, J. R., Beyreuther, K., Kaptein, R., Scheek, R. M. and Hull, W. E. (1983). *Eur. J. Biochem.*, **132**, 321–327

Caruthers, M. H. (1980). *Acc. Chem. Res.*, **13**, 155–160

Clore, G. M., Gronenborn, A. M., Brünger, A. T. and Karplus, M. (1985). *J. Mol. Biol.*, **186**, 435–455

Culard, F., Schnarr, M. and Maurizot, J. C. (1982). *EMBO J.*, **1**, 1405–1409

De Vlieg, J., Boelens, R., Scheek, R. M., Kaptein, R. and van Gunsteren, W. F. (1986). *Israel J. Chem.*, **27**, 181–188

Ebright, R. H., Cossart, P., Gicquel-Sanzey, B. and Beckwith, J. (1984). *Nature (London)*, **311**, 232–235

Ebright, R. H. (1985). *J. Biomolec. Struct. Dyn.*, **3**, 281–297

Ernst, R. R., Bodenhausen, G. and Wokaun, A. (1987). *Principles of Nuclear Magnetic Resonance in One and Two Dimensions*. Clarendon Press, Oxford

Geisler, N. and Weber, K. (1977). *Biochemistry*, **16**, 938–943

Gilbert, W., Gralla, J., Majors, J. and Maxam, A. (1976). In *Protein–Ligand Interactions*, Sund, H. and Blauer, G. (eds), de Gruyter, Berlin, 193–210

Gilbert, W. and Maxam, A. (1973). *Proc. Natl Acad. Sci. USA*, **70**, 3581–3584

Gilbert, W. and Müller-Hill, B. (1966). *Proc. Natl Acad. Sci. USA*, **56**, 1891–1898

Hare, D. R., Wemmer, D. E., Chou, S. H., Drobny, G. H. and Reid, B. R. (1983). *J. Mol. Biol.*, **171**, 319–336

Havel, T. F., Crippen, G. M. and Kuntz, I. D. (1979). *Biopolymers* **18**, 73–81

Havel, T. F., Kuntz, I. D. and Crippen, G. M. (1983). *Bull. Math. Biol.*, **45**, 665–720
Havel, T. F. and Wüthrich, K. (1985). *J. Mol. Biol.*, **182** 281–294
Hochschild, A., Douhan, J. and Ptashne, M. (1986). *Cell*, **47**, 807–816
Hochschild, A. and Ptashne, M. (1986). *Cell*, **44**, 925–933
Jacob, F. and Monod, J. (1961). *J. Mol. Biol.*, **3**, 318–353
Jarema, M. C., Lu, P. and Miller, J. H. (1981). *Proc. Natl Acad. Sci. USA*, **78**, 2707–2711
Jeener, J., Meier, B. H., Backmann, P. and Ernst, R. R. (1979). *J. Chem. Phys.*, **71**, 4546–4553
Kania, J. and Brown, D. T. (1976). *Proc. Natl Acad. Sci. USA*, **73**, 3529–3533
Kaptein, R. (1982). In *Biological Magnetic Resonance*, Berliner, L. J. and Reuben, J. (eds), Plenum, New York, Vol. 4, 145–191
Kaptein, R., Boelens, R., Scheek, R. M. and van Gunsteren, W. F. (1988). *Biochemistry*, **27**, 5389–5395
Kaptein, R., Scheek, R. M., Zuiderweg, E. R. P., Boelens, R., Klappe, K. J. M., van Boom, J. H., Rüterjans, H. and Beyreuther, K. (1983). In *Structure and Dynamics: Nucleic Acids and Proteins*, Clementi, E. and Sarma, R. H. (eds), Adenine Press, New York, 209–225
Kaptein, R., Zuiderweg, E. R. P., Scheek, R. M., Boelens, R. and van Gunsteren, W. F. (1985). *J. Mol. Biol.*, **182**, 179–182
Karplus, M. (1959). *J. Chem. Phys.*, **30**, 11–15
Lehming, N., Sartorius, J., Niemöller, M., Genenger, G., von Wilcken-Bergmann, B. and Müller-Hill, B. (1987). *EMBO J.*, **6**, 3145–3153
Lu, P., Cheung, S. and Arndt, K. (1983). *J. Biomol. Struct. Dyn.*, **1**, 509–521
McCammon, J. A. and Harvey, S. C. (1987). *Dynamics of Proteins and Nucleic Acids*, Cambridge University Press, Cambridge
McKay, D. and Steitz, T. (1981). *Nature*, **290**, 744–749
Matthews, B. W., Ohlendorf, D. H., Anderson, W. F. and Takeda, Y. (1982). *Proc. Natl Acad. Sci. USA*, **79**, 1428–1452
Miller, J. H. (1979). *J. Mol. Biol.*, **131**, 249–258
Miller, J. H. (1984). *J. Mol. Biol.*, **180**, 205–212
Miller, J. H. and Reznikoff, W. (1978). *The Operon*, 2nd edn, Cold Spring Harbor Press, New York
Nick, H., Arndt, H., Boschelli, F., Jarema, M. C., Lillis, M., Sadler, J., Caruthers, M. and Lu, P. (1982). *Proc. Natl Acad. Sci. USA*, **79**, 218–222
Ogata, R. T. and Gilbert, W. (1979). *J. Mol. Biol.*, **132**, 709–728
Ohlendorf, D. H., Anderson, W. F., Fischer, R. G., Takeda, Y. and Matthews, B. (1982). *Nature (London)*, **298**, 718–723
Pabo, C. and Lewis, M. (1982). *Nature (London)*, **298**, 443–447
Pabo, C. and Sauer, R. (1984). *Ann. Rev. Biochem.*, **53**, 293–321
Ribeiro, A. A., Wemmer, D., Bray, R. P., Wade-Jardetzky, N. G. and Jardetzky, O. (1981). *Biochemistry*, **20**, 818–823
Sadler, J. R., Sasmor, H. and Betz, J. L. (1983). *Proc. Natl Acad. Sci. USA*, **80**, 6785–6789
Sauer, R. T., Yocum, R. R., Doolittle, R. F., Lewis, M. and Pabo, C. O. (1982). *Nature (London)*, **298**, 447–451
Scheek, R. M., Russo, N., Boelens, R. and Kaptein, R. (1983a). *J. Am. Chem. Soc.*, **105**, 2914–2916
Scheek, R. M., Zuiderweg, E. R. P., Klappe, K. J. M., van Boom, J. H., Kaptein, R., Rüterjans, H. and Beyreuther, K. (1983b). *Biochemistry*, **22**, 228–235
Scheek, R. M., Boelens, R., Russo, N., van Boom, J. H. and Kaptein, R. (1984). *Biochemistry*, **23**, 1371–1376
Scheek, R. M., Boelens, R., Russo, N. and Kaptein, R. (1985). In *Structure and Motion: Membranes, Nucleic Acids and Proteins*, Clementi, E., Corongiu, G., Sarma, M. H. and Sarma, R. H. (eds), Adenine Press, Guilderland, 485–495
Scheek, R. M. and Kaptein, R. (1989). In *NMR in Enzymology*, Oppenheimer, N. J. and James, T. L. (eds), Academic Press, New York, in the press
Schevitz, R. G., Otwinowski, Z., Joachimiak, A., Lawson, C. L. and Sigler, P. B. (1985). *Nature (London)*, **317**, 782–786
Simons, A., Tils, D., von Wilcken-Bergmann, B. and Müller-Hill, B. (1984). *Proc. Natl Acad. Sci. USA*, **81**, 1624–1628

Steitz, T. A., Ohlendorf, D. H., McKay, D. B., Anderson, W. F. and Matthews, B. W. (1982). *Proc. Natl Acad. Sci. USA*, **79**, 3097–3100

Van Gunsteren, W. F., Kaptein, R. and Zuiderweg, E. R. P. (1983). In *Nucleic Acid Conformation and Dynamics*, Olson, W. K. (ed.), Report of Nato/CECAM Workshop, Orsay, 79–92

Wagner, G., Braun, W., Havel, T. F., Schaumann, T., Go, N. and Wüthrich, K. (1987). *J. Mol. Biol.*, **196**, 611–639

Weber, I. T., McKay, D. B. M. and Steitz, T. A. (1982). *Nucl. Acids. Res.*, **10**, 5085–5102

Wharton, P. P. and Ptashne, M. (1985). *Nature*, **316**, 601–605

Wharton, P. P. and Ptashne, M. (1987). *Nature (London)*, **326**, 888–891

Wüthrich, K. (1986). *NMR of Proteins and Nucleic Acids*, Wiley, New York

Zuiderweg, E. R. P., Scheek, R. M., Veeneman, G., Kaptein, R., Rüterjans, H. and Beyreuther, K. (1981). *Nucl. Acids. Res.*, **9**, 6553–6569

Zuiderweg, E. R. P., Kaptein, R. and Wüthrich, K. (1983a). *Eur. J. Biochem.*, **137**, 279–292

Zuiderweg, E. R. P., Kaptein, R. and Wüthrich, K. (1983b). *Proc. Natl Acad. Sci. USA*, **80**, 5837–5841

Zuiderweg, E. R. P., Scheek, R. M. and Kaptein, R. (1985a). *Biopolymers*, **24**, 2257–2277

Zuiderweg, E. R. P., Scheek, R. M., Boelens, R., van Gunsteren, W. F. and Kaptein, R. (1985b). *Biochimie*, **67**, 707–715

4

The single-stranded DNA binding protein of *Escherichia coli*: physicochemical properties and biological functions

Joachim Greipel, Claus Urbanke and Günter Maass

INTRODUCTION

Single-stranded DNA binding proteins fulfil important functions in DNA metabolism. They have been shown to be essential for replication, recombination and repair in bacteria and bacteriophages. The best-studied single-stranded DNA binding proteins are the gene32 protein from T4-phage (gp32) (for reviews, cf. Kowalczykowski *et al.*, 1981; Chase and Williams, 1986; Chase, 1984), the gene5 protein from filamentous phages (gp5) (for reviews, cf. Kowalczykowski *et al.*, 1981), and the *E. coli* single-stranded DNA binding protein.

Since physicochemical investigations have been a major source of information concerning single-stranded DNA binding proteins and their interactions with DNA and other proteins, we concentrate in this review on physicochemical aspects. Doing this, we intend not only to summarize the most recent investigations in the field but also to reconsider older findings in the light of present knowledge.

We restrict ourselves to the properties and functions of the single-stranded DNA binding protein from *E. coli* (EcoSSB). In contrast to the phage SSBs, which exert very specialized functions in their respective viral replication systems, EcoSSB may constitute a general principle of DNA metabolism, at least in procaryotes.

Although the role of single-stranded DNA binding proteins in procaryotic systems has been well established, there is, in comparison, little information on functionally similar proteins in eucaryotes. The best-studied protein so far among the eucaryotic proteins has been isolated from adenovirus (Field *et al.*, 1984; VanAmerongen *et al.*, 1987). Other proteins with ss-DNA binding capabilities have been isolated from yeast, calf thymus and rat (Chase and Williams, 1986). Recently, it could be shown, however, that proteins which bind to ss-DNA, and which stimulate their respective polymerases, are not necessarily involved in DNA metabolism (Richter *et al.*, 1986). Several of these eucaryotic single-stranded DNA binding proteins could be identified as previously known proteins, e.g. as dehydrogenases (Perucho *et al.*, 1977; Williams *et al.*, 1985; Grosse *et al.*, 1986), thymidylate-kinase (Jong and Campbell, 1984), and an hnRNP (Valentini *et al.*, 1985).

PHYSICAL PROPERTIES AND STRUCTURE OF EcoSSB

EcoSSB is a tetrameric protein of identical subunits (Bandyopadhyay and Wu, 1978; Williams *et al.*, 1983; Molineux *et al.*, 1974; Weiner *et al.*, 1975). The relative molecular mass of the protomer is 18 843, as deduced from the sequence of the ssb gene (Sancar *et al.*, 1981; Chase *et al.*, 1983). Table 4.1 lists hydrodynamic properties of EcoSSB.

The isoelectric point of EcoSSB is 6.0 (Weiner *et al.*, 1975; Williams *et al.*, 1983). The solubility of EcoSSB is relatively low and depends upon salt conditions (Schomburg, 1984) (Table 4.2). EcoSSB contains no cysteine residues.

Table 4.1   Hydrodynamic properties of EcoSSB

Sedimentation coefficient, $s_{20,w}$	4.9^a
	4.4^b
	4.6^c
Frictional ratio, f/f_0	1.36^a
	1.42^c
Stokes radius, r_S (nm)	3.8^a
	3.9^c
Diffusion coefficient, D (10^{-12} m^2 s^{-1})	56^a
	55 ± 3^d

[a] Weiner *et al.* (1975).
[b] Krauss *et al.* (1981).
[c] Williams *et al.* (1983).
[d] Own observations from quasielastic light scattering.

Table 4.2   Solubility of EcoSSB

c(NaCl)	c(EcoSSB)$_{max}$
50 mM	0.5 mg/ml
200 mM	1.5 mg/ml
500 mM	10.0 mg/ml

Buffer: 20 mM K-phosphate, pH 7.5, 4 °C.

The rotational correlation time, τ_c, of the EcoSSB tetramer has been determined from the fluorescence anisotropy decay curve of labelled

EcoSSB to be 108 ns (Bandyopadhyay and Wu, 1978). This value is in good agreement with the results of a recent NMR study, where a τ_c of 80 ns for an EcoSSB–oligonucleotide complex was obtained (Clore *et al.*, 1986). These rotational correlation times are consistent with the reported Stokes radius r_S of approximately 3.8 nm (Weiner *et al.*, 1975). The theoretical values for τ_c and r_S of an unhydrated spherical protein with a molecular mass of 75 400 are 22 ns and 2.8 nm, respectively. The high discrepancy between the observed and the calculated values indicates an unusually high amount of hydration of EcoSSB and/or a considerable deviation from the spherical shape. In electron microscopic studies, the EcoSSB tetramer appeared as an approximately globular particle, with a diameter of 6 ± 0.5 nm (Greipel *et al.*, 1987).

Several crystallizations of EcoSSB have been reported (Ollis *et al.*, 1983; Monzingo and Christiansen, 1983; Hilgenfeld *et al.*, 1984). Preliminary X-ray studies revealed the usual D_2-symmetry for the EcoSSB tetramer (Ollis *et al.*, 1983).

The secondary structure of EcoSSB has been predicted from the amino acid sequence (Sancar *et al.*, 1981), using the method of Chou and Fasman. Following this prediction the 105 aminoterminal amino acids make up a region consisting of approximately equal amounts of alpha-helix and beta-sheets and less than 20 per cent random coil. For the residues 105 to 166, no secondary structure has been proposed. This part consists of more than 80 per cent proline, glycine, asparagine and glutamine, and contains only two charged residues. The remaining 11 carboxyterminal amino acids were predicted to adopt an alpha-helical structure. Five negatively charged amino acid side chains are located in this region.

Further information about the structure of EcoSSB are inferred from proteolysis studies (Williams *et al.*, 1983). The carboxyterminal part of EcoSSB can easily be cleaved off by chymotrypsin (at residue 135) and trypsin (at residue 115). Starting from the beginning of the 'random coil region' at residue 105, these cleavage sites are the first sites to match the sequence requirements of the respective protease. The resulting amino-terminal 'core' proteins bind to DNA with higher affinities than native EcoSSB. The carboxyterminal part is particularly sensitive to proteolytic cleavage if the protein is bound to single-stranded polynucleotides, whereas short oligonucleotides do not affect the cleavage rate.

Recently, the construction of a fusion protein consisting of β-galactosidase, a collagenase recognition sequence, and EcoSSB was described by Scholtissek and Grosse (1988). Collagenase treatment of the fusion protein released a modified EcoSSB, which contained an additional 40 amino acids at the N-terminus. This protein proved to bind to M13-phage single-stranded DNA as tightly as native EcoSSB, and also adopted a tetrameric structure in solution. These results suggest that the amino-terminus of EcoSSB must be located near the protein surface at a site distant from the sites of DNA binding and tetramerization of the EcoSSB subunits.

EcoSSB contains 4 Trp and 4 Tyr residues per monomer (Sancar *et al.*, 1981). Only the Trp residues are responsible for the fluorescence emission of EcoSSB. The fluorescence decay of EcoSSB shows two characteristic lifetimes of 4.1 and 13.2 ns, indicating a heterogeneity of the microenvironment of the EcoSSB tryptophan residues. The high accessibility of the tryptophan residues to quenchers indicates an exposure of the fluorophores on the protein surface (Bandyopadhyay and Wu, 1978).

PURIFICATION OF EcoSSB

Any search for an unknown single-stranded DNA binding protein initially contains a basic purification concept first outlined by Alberts *et al.* (1968), in which immobilized single-stranded DNA is used for the affinity chromatography of DNA binding proteins. A crude cell extract is applied to a single-stranded DNA cellulose column at low ionic strength. Most proteins that are adsorbed on the column can be eluted using the polyanion dextran-sulphate and increasing NaCl concentrations. The protein fraction that elutes at $c_{NaCl} > 1$ M is considered to contain the single-stranded DNA binding proteins.

Generally, it is a non-trivial task to prove that the binding of single-stranded DNA *in vitro* is a physiological function of the isolated proteins *in vivo*. Two 'single-stranded DNA binding' proteins isolated from eucaryotes using similar procedures have been identified as lactate-dehydrogenase (Williams *et al.*, 1985) and glyceraldehyde-3-phosphate dehydrogenase (Perucho *et al.*, 1977; Grosse *et al.*, 1986). The well-investigated single-stranded DNA binding protein UP1 from calf thymus is presumably a degradation product of an hnRNP protein (Williams *et al.*, 1985).

Sigal *et al.* (1972) discovered EcoSSB, applying the basic procedure of Alberts to extracts of *E. coli*. Many of the protocols for the purification of EcoSSB currently in use have evolved from this scheme. Some selected properties of EcoSSB useful for purification are, in brief:

- EcoSSB elutes from single-stranded DNA cellulose columns at NaCl concentrations between 1 and 2 M (Sigal *et al.*, 1972).
- The solubility of EcoSSB is strongly dependent on the temperature and on the concentrations of glycerol and NaCl. At 4 °C and $c_{NaCl} = 50$ mM, the solubility of EcoSSB is approximately 0.5 mg/ml. In a storage buffer containing 1 M NaCl and 60 per cent glycerol, EcoSSB concentrations of 20 mg/ml can easily be obtained (Schomburg, 1985).
- The aggregation of EcoSSB at low temperatures and low NaCl concentrations is reversible if the solution contains glycerol as a stabilizing

agent. Precipitation of EcoSSB by dialysis against potassium phosphate buffers of low ionic strength was originally introduced as a purification step by Sigal *et al.* (1972) and is still in use (Chase *et al.*, 1984). This precipitated EcoSSB redissolves upon dialysis against 10 per cent glycerol containing no salts at all. We observed that the redissolved EcoSSB again precipitates upon adding minute amounts of potassium phosphate (unpublished).

– The fact that EcoSSB does not precipitate upon heating to 100 °C for 2 minutes has been used as a purification step (Weiner *et al.*, 1975; Krauss *et al.*, 1981).

– EcoSSB is precipitated by relatively low ammonium sulphate concentrations ($\approx$27 per cent saturation). This property of EcoSSB has been used for purification (Meyer *et al.*, 1980; Lohman *et al.*, 1986a).

– EcoSSB binds to the dye Cibacron F3G immobilized for affinity chromatography of EcoSSB. Meyer *et al.* (1980) coupled the commercially available soluble Cibacron F3G containing polymer 'Blue-dextran' to Sepharose. EcoSSB can be eluted from the resulting material at $c_{NaCl} = 2$ M ('Blue-dextran-Sepharose'). Krauss *et al.* (1981) used a method by Böhme *et al.* (1972) for a direct coupling of Cibacron F3G to Sepharose CL-4B. The resulting material ('Blue-Sepharose') binds EcoSSB much more tightly, so 2 M NaCl and 5 M urea are needed for the elution.

– EcoSSB elutes from DEAE-based anion-exchangers at $c_{NaCl} = 200$ mM at pH 8 (Sigal *et al.*, 1972).

– Further purification protocols for EcoSSB include the use of heparin-sepharose (Krauss *et al.*, 1981), hydroxyapatite (Bobst *et al.*, 1985), chromatography on PBE-94 (Williams *et al.*, 1984) and chromatography on phosphocellulose (Weiner *et al.*, 1975).

The straightforward purification of milligram amounts of EcoSSB has become possible with the application of cloning techniques. The presence of the multi-copy plasmids pDR2000 (Chase *et al.*, 1980) and pDR1996 (Williams *et al.*, 1983) in *E. coli* lead to an approximately ten- to twentyfold overproduction of EcoSSB as compared with the parent wildtype strain. The overproduction of EcoSSB does not exert lethal effects on the bacterial cell.

To date, many groups use temperature-inducible expression vectors in which the gene of wild-type or mutant EcoSSB is under control of the P_L-promoter of the bacteriophage λ (Lohman *et al.*, 1986a; Khamis *et al.*, 1987c; Bayer *et al.*, 1989). After some hours of induction, EcoSSB represents up to 10 per cent of the total cell protein. Purification of EcoSSB from these overproducing strains easily leads to final yields of more than 5 mg EcoSSB per g of wet cells (Lohman *et al.*, 1986a). Two different protocols for the purification of large amounts of mutant and wild-type EcoSSB from temperature-inducible overproducers have been described by Williams *et al.* (1984) and Lohman *et al.* (1986a).

CONCENTRATION DETERMINATION OF EcoSSB

The accurate determination of EcoSSB concentrations is crucial in all experiments concerning the stoichiometry of EcoSSB–DNA interaction. Determinations of protein concentration by UV absorption depend critically on the exact knowledge of the extinction coefficient ε.

Considerably different values for $\varepsilon^{280\,nm}$ have been reported using various methods. Ruyechan and Wetmur (1976) determined $\varepsilon = 113\,000$ $M(Tetr.)^{-1}$ cm^{-1} (corrected for a molecular weight of 18857 per monomer), using colorimetric methods (Lowry, Biuret). Amino acid analysis yielded $\varepsilon = 121\,000$ $M(Tetr.)^{-1}$ cm^{-1} (Williams *et al.*, 1983; Lohman and Overman, 1985). Williams *et al.* (1984) calculated $\varepsilon = 108\,000$ $M(Tetr.)^{-1}$ cm^{-1} from the known amino-acid composition of EcoSSB. We obtained $\varepsilon = 94\,800$ $M(Tetr.)^{-1}$ cm^{-1} from measurements of the refractive index increment (Babul and Stellwagen, 1969). This value was also used by Krauss *et al.* (1981), although a different extinction coefficient was published there, owing to an error in the preparation of the manuscript.

STRUCTURAL FEATURES OF EcoSSB–DNA COMPLEXES

Complexes of EcoSSB with single-stranded nucleic acids have been visualized by electron microscopy. The results, however, are difficult to compare, since the conditions for sample preparation, as well as the fixing and staining procedures, differ considerably.

Sigal *et al.* (1972) visualized a complex of EcoSSB and circular single-stranded DNA from fd-Phage by negative staining. The EcoSSB-covered DNA appeared as a 'beaded necklace' of about 200 beads, a value that leads to a stoichiometry of about 33 nucleotides per bead. The spacing of the beads was about 6 nm. The very low length increment of 0.18 nm per nucleotide in the fd-DNA–EcoSSB complex led to the conclusion that the DNA must be in a 'regularly folded' conformation; no influence of spermine on the structure of the complex was observed.

In a recent study, a beaded nucleosome-like structure was also observed for complexes of EcoSSB and poly(dT) at an NaCl concentration of 300 mM (Greipel *et al.*, 1987). Shape and size of the observed DNA-bound protein particles and the excess free EcoSSB tetramers were identical. Diameter and spacing of the beads, as well as the length increment of 0.18 nm per nucleotide, were comparable to the results of Sigal *et al.* (1972). At an excess of nucleic acid, the EcoSSB particles were distributed randomly between the poly(dT) strands. EcoSSB-saturated nucleic acid strands adopted a rather inflexible wormlike structure, compared with the coiled structure of the partly covered poly(dT) molecules.

A somewhat different description of the structure of EcoSSB–nucleic acid complexes has been given by Chrysogelos and Griffith (1982, 1984). These authors also report an organization of circular, single-stranded fd-DNA in a nucleosome-like structure. In contrast to the results of Sigal *et al.* (1972), however, they observe an average number of 38 to 40 beads per fd-DNA circle. This number corresponds to a stoichiometry of about 160 bases per bead. The contour length of a fully complexed fd-DNA circle was reported to be about 540 nm, corresponding to a length increment of 0.8 nm per base. The average bead diameter of 12 nm also differs considerably from the value of 5 to 6 nm obtained by Sigal *et al.* (1972) and Greipel *et al.* (1987). At protein to DNA concentration ratios above 6:1 (weight/weight), Chrysogelos and Griffith (1984) report a replacement of the beaded structure by a more smoothly contoured form with almost doubled contour length. From equilibrium density gradient centrifugation and nuclease digestion of EcoSSB–DNA complexes, Chrysogelos and Griffith deduced that EcoSSB binds to DNA as an octameric particle covering 145 to 170 bases. A similar digestion experiment has been reported by Boidot-Forget *et al.* (1986). When complexes of poly(dT) and EcoSSB were digested with P1-nuclease, a pattern of DNA fragments was observed with a maximum intensity at a fragment length of 80 nucleotides.

Further structural information can be obtained from a number of electron microscopic studies that did not focus primarily on the fine structure of the EcoSSB–DNA complex, but rather on the kinetics and thermodynamics of the EcoSSB–DNA interaction or the interaction of EcoSSB with other proteins: Ruyechan and Wetmur (1975) investigated the length distribution of complexes formed between denatured DNA of Lambda-phage and EcoSSB as a function of salt concentration at different protein-to-DNA ratios. They observed extended EcoSSB covered regions as well as 'collapsed irregular' uncovered regions on the same DNA molecule.

Schneider and Wetmur (1982) visualized complexes between EcoSSB and circular DNA from G4-phage in the electron microscope. They observed mainly protein-free or completely covered circular single-stranded DNA. The fully complexed G4-DNA appeared 'round and large', whereas partially covered DNA looked 'folded or twisted with one or more extended loops'.

Electron micrographs in which DNA covered by RecA protein and EcoSSB is visualized have been obtained by Register and Griffith (1985) and by Williams and Spengler (1986). Their data also suggest a beaded, nucleosome-like structure for EcoSSB–DNA complexes.

A major problem in investigating EcoSSB–nucleic acid complexes arises from the very badly defined structure of the free single-stranded DNA. In contrast to double-stranded DNA, where the B-type helix is the predominant form, at most solution conditions no general structure can be given for single-stranded nucleic acids. The often-used single-stranded phage DNAs, e.g., are well known to form secondary and tertiary structures, which could lower the accessibility of the DNA to EcoSSB (Reckmann *et al.*, 1985). Unfortunately, there is no electron microscopic technique available that will give images of a free, unperturbed, single-stranded nucleic acid. The contour lengths of EcoSSB–nucleic acid complexes observed in the electron microscope can therefore not easily be compared with the contour lengths of the uncovered nucleic acid, simply because the latter is not known. The low spacing of 0.18 nm per nucleotide for EcoSSB–nucleic acid complexes supports, however, the view that the DNA is coiled around the EcoSSB tetramer. This conclusion was also drawn by Krauss *et al.* (1981) from thermodynamic data.

Since no X-ray structure of EcoSSB or its complexes with DNA is available, only few data concerning the structure of the EcoSSB–DNA complexes and the nature of the EcoSSB–DNA interactions are known. Bandyopadhyay and Wu (1978) concluded, from chemical modification of the EcoSSB protein, that tryptophan and lysine residues are involved in the protein–DNA interaction, whereas tyrosine and arginine are not.

Merrill *et al.* (1984) showed that the residue Phe-60 of the EcoSSB can be photo-crosslinked to DNA. The interaction of tryptophan residues of EcoSSB with the bases of the DNA has been extensively studied by means of optically detected magnetic resonance (Cha and Maki, 1984; Khamis *et al.*, 1987a, b, c, d; Casas-Finet *et al.*, 1987a, b, 1988; Zang *et al.*, 1988) and site-directed mutagenesis (Khamis *et al.*, 1987c; Casas-Finet *et al.*, 1988; Zang *et al.*, 1988). These authors conclude that there exists a stacking interaction between a thymine base of poly(dT) and the aromatic residues Trp-54 and Phe-60. The residue Trp-135 was shown not to participate in DNA binding.

Further insight into the structure of EcoSSB–DNA complexes comes from NMR studies. In a 270 MHz study, Römer *et al.* (1984) did not observe high-field shifts exceeding 0.05 p.p.m. when a complex between poly(dT) and EcoSSB was formed, and concluded the absence of 'extensive stacking interactions'. In the same work, it was also shown that the resonances of poly(dT) are broadened beyond detection when the polynucleotide is more than about 45 per cent saturated with EcoSSB. This broadening reflects the low rotational mobility of the nucleotides in the complex with EcoSSB.

The single-stranded deoxy-oligonucleotide 5'-AAGTGTGA TAT-3' is known to adopt a right-handed helical structure in solution (Clore and Gronenborn, 1984). In a 500 MHz NMR study, using the transferred nuclear Overhauser enhancement, it was shown that the overall structure of the oligomer is not disturbed upon binding to EcoSSB. Additionally, it was observed that in the complex the bases of the DNA oligomer undergo considerably slower rotational motions than the sugar moieties. This finding supports the view that hydrophobic interaction of the bases of the DNA with aromatic sidechains of EcoSSB are a major cause of complex stability.

INTERACTIONS OF *E. coli* SINGLE-STRANDED DNA BINDING PROTEIN WITH OTHER PROTEINS

There have been numerous reports on the influence of EcoSSB on the function of other proteins involved in interactions with nucleic acids. The most thoroughly investigated protein influenced by EcoSSB has been the RecA gene product. RecA protein plays an important role in the recombination and repair of DNA (Walker, 1984). It is one of the first proteins induced in SOS repair. The effects of EcoSSB on the SOS functions have been shown by mutations. The temperature-sensitive mutations ssb-1 (Meyer *et al.*, 1979) and ssb-113 (Chase *et al.*, 1984) are defective in a number of SOS responses. Both are defective in the induction of λ prophage and amplification of RecA protein synthesis after UV irradiation or mitomycin treatment (Baluch *et al.*, 1980). For the ssb-1 mutant, these defects are expressed only at the non-permissive temperature, whereas the ssb-113 mutant is defective at both the permissive and non-permissive temperatures (30 ° and 42 °, resp.).

Most of the functions of RecA are concerned with interactions between the protein and single-stranded or double-stranded DNA. *In vitro* RecA catalyses the homologous pairing of single-stranded DNA with double-stranded DNA, and thus the formation of D-loops in the presence of ATP (Radding *et al.*, 1982). EcoSSB accelerates this reaction considerably (Riddles and Lehman, 1985a, b; McEntee *et al.*, 1980; Cox *et al.*, 1983; Cassuto *et al.*, 1980; West *et al.*, 1982). The effect has been attributed to the destabilization of DNA secondary structures by EcoSSB. Electron microscopic analysis showed that in the initial stages of D-loop formation, the single-stranded DNA is covered by EcoSSB (Register and Griffith, 1985). The protein then is replaced by RecA until no EcoSSB remains on the single-stranded circular DNA, whereas some EcoSSB still remains at the 5'-end of a linearized single-stranded DNA (Register and Griffith, 1985). It is concluded that RecA assembles on the single-stranded DNA in a unidirectional manner from the 5'- to the 3'-end. Further events in D-loop formation seem to be independent of EcoSSB and single-stranded DNA binding proteins from phages and plasmids (Egner *et al.*, 1987).

However, a stable ternary complex of EcoSSB, RecA and single-stranded DNA has been described using as an indicator the quench of the fluorescence of EcoSSB bound to single-stranded DNA (Morrical *et al.*, 1986). This complex can be formed only by adding EcoSSB to a preformed stable complex of RecA and single-stranded DNA in the presence of ATP indicating a kinetic stabilization. In the absence of ATP, RecA competes with EcoSSB for binding to the single-stranded DNA in a normal manner.

In recent investigations, an attempt was made to rationalize all effects of EcoSSB on RecA in one model (Kowalczykowski *et al.*, 1987; Kowalczykowski and Rupp, 1987). In this model, EcoSSB competes with RecA for binding to single-stranded DNA. In the absence of ATP or analogues of ATP, EcoSSB is the more potent ligand, whereas, in the presence of ATP, RecA can displace EcoSSB from single-stranded DNA. Intermediate stages can be observed. The effect of EcoSSB on RecA is to facilitate binding by removing secondary structure from the DNA template. In addition, a complex of RecA and DNA containing double-stranded regions was observed that could not be replaced by EcoSSB. Again, no protein–protein interaction of EcoSSB with RecA could be demonstrated.

In prophage induction, RecA protein has a proteolytic activity towards the phage λ repressor. This proteolysis is dependent upon the presence of single-stranded DNA, and is enhanced by EcoSSB as well as by the ssb-113 mutant protein (Resnick and Sussman, 1982). Both proteins relieved the inhibition of proteolysis by an excess of denatured DNA. A large excess of ssb-113 protein, however, also inhibited proteolysis, whereas the wild-type EcoSSB did not.

RecA also catalyses the reannealing of single-stranded DNA. This reaction is inhibited by EcoSSB (McEntee, 1985; Cohen *et al.*, 1983), owing possibly to the tight binding of the protein to the single-stranded DNA.

The ability to enhance the activity of DNA polymerase was one of the first tests used to identify single-stranded DNA binding proteins (Sigal *et al.*, 1972). Binding of EcoSSB to the template enhances the accuracy of several DNA polymerases from prokaryotes and eukaryotes up to tenfold (Kunkel *et al.*, 1979). *In vitro* studies showed a large influence of EcoSSB on *E. coli* DNA polymerase III replicating single-stranded phage DNA. The polymerase holoenzyme was activated by the presence of EcoSSB, and the processivity was enlarged to such an extent that the polymerase could replicate up to 3000 nucleotides in one binding event. In contrast, the core enzyme was almost completely inhibited by EcoSSB (Fay *et al.*, 1981). EcoSSB could not enhance the activity of the adenovirus DNA polymerase, whereas the adenovirus DNA binding protein could (Field *et al.*, 1984).

It was also shown that EcoSSB inhibits the transcription on single-stranded phage M13 DNA by *E. coli* RNA polymerase (Niyogi *et al.*, 1977). It was shown that EcoSSB can also act as part of an isolatable pre-priming complex at the *E. coli* origin of replication (VanDerEnde *et al.*, 1985).

In ΦX174 replication, gene A product acts as a site-specific endonuclease, cleaving in the origin of replication. In absence of EcoSSB, the specificity of the nuclease is relaxed, resulting in the appearance of another nucleolytic cleavage at a site homologous to the origin (VanMansfeld *et al.*, 1986). In this case, EcoSSB protects the DNA, except the unique origin structure. A direct protein–protein interaction seems 'not plausible'.

There is very little indication of a direct interaction of EcoSSB with other proteins. Direct interactions between EcoSSB and polymerase II and exonuclease I of *E. coli* have been indicated by early co-sedimentation experiments (Molineux and Gefter, 1974, 1975). Recently, a report has been given on the retention of several proteins on EcoSSB affinity columns (Perrino *et al.*, 1986). One of these retarded proteins is thought to be a subunit of *E. coli* DNA polymerase III.

As a result of these numerous investigations on the interaction of EcoSSB with other proteins, it can be concluded that there is strong evidence for a model in which all of the interactions are brought about by the binding of EcoSSB to single-stranded DNA. An active complex of EcoSSB with another protein could not be demonstrated and could be ruled out in some cases.

INTERACTIONS OF EcoSSB WITH NUCLEIC ACIDS

Single-stranded DNA binding proteins have been functionally defined as those nucleic acid binding proteins that bind preferentially to single strands. This property was employed in the first preparation of EcoSSB (Sigal *et al.*, 1972). In this study, it was also shown that EcoSSB has very little or no affinity towards double-stranded DNA. The affinity of EcoSSB towards single-stranded nucleic acids is modulated to a large extent by the nature of the nucleic acid employed.

Methods to detect binding

Co-sedimentation studies showed that EcoSSB binds strongly to fd-DNA, but only weakly or not at all to R17 RNA (Sigal *et al.*, 1972). In an analogous approach, the binding of EcoSSB to the oligonucleotides $d(pT)_8$, $d(pT)_{16}$ and $d(pT)_{30-40}$, as well as fd-DNA, was demonstrated (Krauss *et al.*, 1981).

Measuring the retention of radioactively labelled EcoSSB on gel filtration columns (Weiner *et al.*, 1975) showed that $\Phi X174$ single-stranded DNA and poly(dT) bind strongly, while poly(dA) and poly(rU) bind only weakly to EcoSSB. For poly(rA), no binding could be detected. On nitrocellulose filters, poly(dC), poly(dT), poly(dI), poly(rU) and poly(rI) could displace $\Phi X174$ single-stranded DNA from the complex with EcoSSB, whereas poly(rC) and poly(rA) could not. Binding to double-stranded DNA could not be detected. Anderson and Coleman (1975) observed binding of EcoSSB to fd single-stranded DNA and poly(dA-dT) by circular dichroism studies. From ESR measurements the binding of EcoSSB to spin-labelled poly(dT) was deduced (Bobst *et al.*, 1985).

Binding of EcoSSB to double-stranded DNAs could only be demonstrated for superhelically stressed DNA (Glikin *et al.*, 1983; Langowski *et al.*, 1985; Srivenugopal and Morris, 1986). The negative superhelical stress of the double-stranded DNA can be relieved by partial melting (Benham, 1985) and EcoSSB can be bound to single-stranded regions of the DNA.

A thermodynamic consequence of the large preference of EcoSSB for single-stranded nucleic acids is the ability of the protein to denature double-stranded DNAs. This effect could be shown for T4 DNA, where at 37 °C EcoSSB could induce a slow, reversible hyperchromic change at 260 nm, and for phage λ DNA, where electron microscopy showed a partial denaturation of the double-strand (Sigal *et al.*, 1972). Williams *et al.* (1983) reported that EcoSSB reduces the melting temperature of poly(dA-dT) by 8 ° when added in stoichiometric amounts. Proteolytic fragments of EcoSSB from which parts of the carboxyterminal end had been removed and the ssb-113 mutant protein reduce the melting temperature even more (Williams *et al.*, 1983; Chase *et al.*, 1984). By denaturing hairpin structures in single-stranded DNA, EcoSSB catalyses the reannealing of complementary DNA strands (Christiansen and Baldwin, 1977).

The binding of several ^{125}J-labelled oligonucleotides could be shown by equilibrium dialysis (Ruyechan and Wetmur, 1976) (Table 4.3).

Binding of EcoSSB to nucleic acids is accompanied by a decrease in protein fluorescence. Most information on the thermodynamics of EcoSSB has been obtained from fluorescence measurements. Binding of EcoSSB could be shown to $d(pT)_{10}$, $d(pT)_{16}$ and fd-DNA (Bandyopadhyay and Wu, 1978), poly(dT), poly(dA), poly(dC), poly(rA), poly(rC), fd-DNA, poly(U), tRNA^{Gln}, $d(pT)_{16}$ and $d(pT)_8$ (Molineux *et al.*, 1975), to poly(rεA) (Lohman, 1986), to poly(rU) and poly(dA) (Lohman *et al.*, 1986b), to fd-DNA, poly(dT), poly(dA), $d(pT)_8$, $d(pT)_{16}$, $d(pT)_{30-40}$ and $d(pA)_{40-60}$ (Krauss *et al.*, 1981), and to poly(dC), poly(dA, dC) and poly(dC, dT) (Schomburg, 1985).

Summarizing these studies, it can be concluded that the affinity of EcoSSB depends on the nature of the nucleic acid employed, and an order of decreasing affinity can be given, even if in most instances no individual binding constants could be derived: poly(dT), phage single-stranded DNA (fd, ΦX174), poly(dC), poly(dC, dT), poly(dA, dC) > poly(rU) ≃ poly(dA) > polyribonucleic acids.

In general, binding of proteins to polyanions like DNA decreases with increasing salt concentration. EcoSSB can be dissociated from nucleic acids by raising the salt concentration above 1 M NaCl. However, it is not possible to dissociate EcoSSB from poly(dT), and natural single-stranded DNAs by addition of NaCl because of its high affinity to these polynuc-

Table 4.3   Binding of short oligonucleotides to EcoSSB

Oligonucleotide	K_{Ass} (M^{-1})	Remarks
d(pCpT)$_2$	0.5–$0.6 \cdot 10^4$	1
d(pCpT)$_3$	3–$5 \cdot 10^4$	1
d(pCpT)$_4$	5–$9 \cdot 10^4$	1
d(pCpT)$_{6-9}$	8–$14 \cdot 10^4$	1
d(pC)$_6$	$5 \cdot 10^4$	1, 2
d(pCpT)$_3$	$5 \cdot 10^4$	1, 2
d(pCpG)$_3$	$4 \cdot 10^4$	1, 2
d(pCpA)$_3$	$2 \cdot 10^4$	1, 2
d(pT)$_8$	$4.7 \cdot 10^3$	3
d(pT)$_6$	$1.4 \cdot 10^3$	3
d(pT)$_5$	$1.3 \cdot 10^3$	3
d(pT)$_4$	$0.9 \cdot 10^3$	3
d(pT)$_3$	$0.5 \cdot 10^3$	3
d(pT)$_2$	$0.4 \cdot 10^3$	3
3'-TMP	$0.3 \cdot 10^3$	3
5'-TMP	—	3
d(pT)$_8$	$2 \cdot 10^4$	4
d(pT)$_{16}$	$6 \cdot 10^5$	4
d(pT)$_{30-40}$	$>3 \cdot 10^8$	4
d(pA)$_{16}$	$5 \cdot 10^3$	4
d(pA)$_{40-60}$	$4 \cdot 10^6$	4

1. 4 °C, 0.2 M NaCl, pH 7.8 (Ruyechan and Wetmur, 1976).
2. Calculated from given $\Delta G°$ values.
3. 22 °C, 0.2 M NaCl, pH 7.4 (Schomburg, 1985).
4. 8 °C, 0.2 M NaCl, pH 7.4 (Krauss *et al.*, 1981).

leotides. For oligonucleotides, it could be shown that binding of EcoSSB to $(dT)_{16}$ is not salt dependent, whereas its affinity to $(dA)_{40-60}$ decreases by two orders of magnitude between 0.05 and 0.4 M KCl (Krauss *et al.*, 1981).

Stoichiometry

The stoichiometry of the binding of EcoSSB has been determined by different methods. From measurements of the helix formation in single-stranded fd-DNA and from co-sedimentation experiments a protein/DNA weight ratio of 8 to 1 in the complex was observed (Sigal *et al.*, 1972), corresponding to the binding of approximately 7 nucleotides per EcoSSB monomer. From nuclease digestion, equilibrium density banding and electron microscopy, it was deduced that 160 ± 25 nucleotides of single-stranded fd-DNA are bound to an octamer of EcoSSB, corresponding to a stoichiometry of 20 nucleotides per EcoSSB monomer (Chrysogelos and Griffith, 1982). In another nuclease digestion experiment, it was found that 80 nucleotides from poly(dT) were covered by one EcoSSB tetramer (Boidot-Forget *et al.*, 1986). From CD titrations, Anderson and Coleman (1975) deduced a stoichiometry of 14 nucleotides of single-stranded fd-DNA bound to one EcoSSB monomer. From UV titrations with poly(dT) and fd-DNA and electron spin resonance measurements with spin-labelled poly(dT), stoichiometries between 6 and 25 nucleotides per EcoSSB monomer were derived, depending on the polymer and the pre-treatment of the EcoSSB (Bobst *et al.*, 1985). The inconsistencies found in the values of these stoichiometries may be due to the large uncertainties inherent in the methods used, but can also reflect the salt dependence of the binding stoichiometry.

A salt dependence of the stoichiometry is observed in fluorimetric titrations of EcoSSB with poly(dT). Lohman and Overman (1985) describe two binding modes: at 25 °C and pH 8.1, the number of T residues covered by one EcoSSB tetramer is 33 ± 3 at low salt ($c_{NaCl} < 0.01$ M) and 65 ± 5 at high salt ($c_{NaCl} > 0.2$ M). In a subsequent study, Bujalowski and Lohman (1986) observe an intermediate mode of binding with 56 ± 3 nucleotides per EcoSSB tetramer at the same conditions. At 37 °C and low salt ($c_{NaCl} = 0.007$–0.012 M), they find a fourth binding mode with 40 ± 2 nucleotides per EcoSSB tetramer. A somewhat different result is obtained from fluorimetric titrations at 8 °C, pH 7.3 and 0.2 M KCl, with approximately 40 nucleotide residues of poly(dT), poly(dA), single-stranded fd-DNA, and poly(rU) covered by one EcoSSB tetramer (Krauss *et al.*, 1981). The same value was also measured by Greipel *et al.* (1987). Schomburg (1985) reports a stoichiometry of 47 ± 3 nucleotides per EcoSSB tetramer for poly(dT), poly(dA) and poly(rU) (cf. Table 4.4). The reason for the differences in the absolute values is not yet clear. It may partially be attributable to differences in the extinction coefficients used for the determination of EcoSSB concentration. Experimental difficulties in the fluorescence titrations could be another reason. Casas-Finet *et al.* (1987a) report that they could reproducibly determine the binding stoichiometry of plasmid-encoded SSB proteins only at $c_{NaCl} > 0.4$ M. Shimamoto *et al.* (1987) describe that adsorption effects can severely disturb the fluorescence titrations with EcoSSB.

The number of nucleotides covered by the EcoSSB tetramer on polymeric lattices must be distinguished from the number of nucleotides actually contributing to the free energy of binding (Kowalczykowski *et al.*, 1981). The latter can be obtained from binding studies on small oligonucleotides. Krauss *et al.* (1981) could show that one molecule of $d(pT)_8$ can bind to each monomeric subunit of EcoSSB, whereas one molecule of $d(pT)_{16}$ spanned two subunits. Going to even smaller oligonucleotides, it could be shown by fluorescence titrations that the smallest unit binding to EcoSSB is a 3'-TMP (cf. Table 4.3) and increasing the chain length up to a $d(pT)_6$ does not change the binding affinity by more than the statistical contribution. A binding of 5'-TMP could not be detected (Schomburg, 1985). This suggests that the monomeric subunit of EcoSSB interacts with only one nucleotide residue and only one phosphate moiety, and that more than eight residues are needed to span the distance between two binding sites. A qualitatively similar result has been obtained by equilibrium

Table 4.4   Binding of EcoSSB to polymeric nucleic acids

Polymer	n	ω	K_i (M^{-1})	T (°C)	NaCl (M)	F_{bound} (%)[a]	Remarks
poly(dT)	47 ± 3	500 ± 200	$2.4 \pm 1 \cdot 10^4$	65	0.2	24 ± 1	1
poly(dT)	47 ± 3	700 ± 300	$21 \pm 1 \cdot 10^4$	55	0.2	20 ± 1	1
poly(dT)	47 ± 3	150 ± 50	$26 \pm 1.5 \cdot 10^4$	55	0.5	20 ± 1	1
poly(dT)	47 ± 3	50 ± 40	$25 \pm 2 \cdot 10^4$	55	0.75	20 ± 1	1
poly(dA)	47 ± 3	500 ± 300	$\approx 1 \cdot 10^4$	22	0.05	37 ± 3	1
poly(dA)	47 ± 3	<30	$\approx 1 \cdot 10^4$	22	0.2	37 ± 3	1
poly(rU)	47 ± 3	100 ± 50	$10 \pm 4 \cdot 10^4$	22	0.05	37 ± 2	1
poly(rU)	47 ± 3	1	$10 \pm 4 \cdot 10^4$	22	0.2	37 ± 2	1
poly(rU)	65	40	$10 \cdot 10^4$	25	0.2	59	2
poly(rU)	65	40–60	1.2–$2.4 \cdot 10^4$	25	0.25	57–62	2, 3
poly(rU)	65	60	$0.6 \cdot 10^4$	25	0.3	57	2, 3
poly(dA)	65	40–60	1.2–$2.2 \cdot 10^4$	25	0.2	54	2
poly(dA)	65	60	$0.7 \cdot 10^4$	25	0.25	54.6	2
poly(dA)	65	40–60	1.5–$2.1 \cdot 10^4$	25	—	78.7	2, 4

[a] Fluorescence of bound EcoSSB compared with free protein.
1. Schomburg (1985); cf. text.
2. Lohman *et al.* (1986b); cf. text.
3. Lohman *et al.* (1986b) report separate evaluations of K_i, F_{bound}, and ω, depending on the concentration of EcoSSB. For reasons of clarity we include only the range of values found.
4. 0.1 M $MgCl_2$ instead of NaCl.

dialysis, although the absolute binding constants are approximately one order of magnitude larger, probably because of different conditions employed (Ruyechan and Wetmur, 1976). For the single-stranded DNA binding protein gp32 from T4 phage, two nucleotide residues contribute to the free energy of binding (Kowalczykowski *et al.*, 1981).

Binding constants

The theoretical basis of non-specific binding of multidentate ligands to linear lattices containing many independent binding sites has been extensively described by McGhee and von Hippel (1974). Closed expressions have been derived for the non-cooperative, as well as for the co-operative, binding of ligands to polymeric lattices of infinite length, which relate the binding density ν (moles of bound ligands per mole of total lattice residue) and the free ligand to the intrinsic association constant K, the ligand site size n, and the co-operativity parameter ω. A similar approach has been chosen by Schwarz (1977; Schwarz and Stankowski, 1979; Schwarz and Watanabe, 1983). The influence of finite lattice length has been numerically evaluated by Epstein (1978). A 'continuum model', where no discrete contact points between the ligand and the polymer are required, has been introduced by Woodbury (1981). The theory of McGhee and von Hippel has been successfully applied to describe the interaction of bacteriophage

T4-coded gene 32 protein with single-stranded nucleic acids (for review, cf. Kowalczykowski *et al.*, 1981). For EcoSSB, these models have so far been used to interpret the binding to low-affinity polynucleotides as poly(dA) and poly(rU) and to poly(dT) at higher temperatures (Lohman *et al.*, 1986b; Schomburg, 1985). For other polymers, such as single-stranded phage DNAs or poly(dT), at room temperature no experimental data are available that would allow the application of these models, because of the high binding affinity, which always yields 'stoichiometric' titrations. Bujalowski and Lohman developed a theory in which the co-operativity of binding of EcoSSB tetramers to single-stranded DNA is limited to the formation of octamers (Bujalowski and Lohman, 1987a, b). They show that fluorimetric titrations of poly(dA) and poly(rU) with EcoSSB are well described by this theory.

The usual way to apply any of these theories to experiments, namely fluorescence titrations, is a non-linear fitting procedure. From the observed fluorescence and the fluorescence of the bound ligand, the concentrations of free and polymer-bound EcoSSB can be calculated. The appropriate theory is used to calculate the concentrations from an assumed intrinsic binding constant K_i, a co-operativity parameter ω, and the binding site size n. Some or all of these parameters are then varied, until the theoretical binding isotherm matches the experimental data. It is important to note that the parameters resulting from such a fitting procedure may depend on each other; e.g., assuming a larger binding constant K_i and at the same time a smaller co-operativity parameter ω will very often not alter the theoretical binding isotherm significantly. In order to obtain satisfying results from such multi-parameter fits, it is important to use a wide concentration range in the titrations. Lohman *et al.* (1986b) provide several examples of such fitting procedures. In Table 4.4, the binding parameters for a variety of polymers are summarized.

The differences found in the interpretations of Lohman *et al.* (1986b) and Schomburg (1985) could arise from the fitting procedures used. Lohman *et al.* (1986b) use a binding site size determined independently by titrations of poly(dT) and determine the fluorescence quench for each polymer and salt concentration from an independent dilution experiment according to Schwarz and Watanabe (1983). They use the theory developed by McGhee and von Hippel (1974) to determine binding constants and co-operativities. On the other hand, Schomburg (1985), employing the theory of Schwarz and Watanabe (1983), fits all four parameters with no independent determinations of the quench and the binding site size.

Overman *et al.* (1988) thoroughly investigated the salt dependence of the binding of EcoSSB to poly(dA), poly(rU), poly(rA) and poly(dT) with different salts under conditions that favour the high-salt binding stoichiometry. They found that the binding co-operativity evaluated for the octamer model of Bujalowski and Lohman (1987a, b) is 400 ± 100 (130 ± 70

for poly(dT)), and does not depend upon the nature and concentration of salt. The binding constant, however, strongly decreases with increasing ionic strength.

KINETICS OF THE BINDING OF EcoSSB TO SINGLE-STRANDED DNAs

EcoSSB functions by binding to single-stranded DNA and preparing the DNA as a substrate for subsequent reactions (e.g. polymerase, RecA). In the course of these reactions, EcoSSB has to leave the complex with single-stranded DNA. The kinetics of interaction of EcoSSB with single-stranded DNAs, therefore, must be considered to get an understanding of the *in vivo* function. For the purpose of this review, we will first concentrate on the theoretical approaches and subsequently discuss the data on the interaction of EcoSSB with polymeric lattices and with short single-stranded oligonucleotides.

EcoSSB binds to the linear single-stranded DNA covering at least 30 nucleotide residues per tetramer, depending on the conditions. Since the initial binding of an EcoSSB molecule can occur at any single nucleotide, the problem of potentially overlapping but mutually exclusive binding sites has to be solved. The first solution to this problem is a pure equilibrium thermodynamical approach. A thorough treatment of the dissociation of co-operatively bound proteins from nucleic acids has been given by Lohman (1983). In this model, dissociation occurs only from the ends of a protein cluster and translocation of ligands on the polymer is excluded. For fully saturated matrices, the dissociation is of zeroth order, whereas for medium saturation 'normal' first-order kinetics are predicted. This theory could be experimentally verified for the dissociation of T4 gene 32 protein from several single-stranded DNAs.

The association of proteins to long linear matrices like DNA has been described in terms of a diffusion-limited reaction. Theories explaining the absolute values of the association rate constants have been developed (for a review, cf. Lohman, 1986, and Berg and von Hippel, 1985). A theory for diffusion-limited association of proteins to globular domains of single-stranded DNAs has been proposed (Mazur and Record, 1986).

The mechanism of association of multidentate ligands to polymeric lattices is very complex. For the reaction of T4 gene 32 protein with single-stranded DNA, a detailed mechanism could be worked out that involves the association of isolated proteins to the DNA, the growth of small clusters of contiguously bound proteins, and, finally, the redistribution of the clusters within the complex (Lohman and Kowalczykowsky, 1981). A similar treatment for the EcoSSB association is still missing. Most association rates reported here have been obtained at a large excess of binding sites over protein, and thus represent the intrinsic association rate constants for the association of a single, isolated EcoSSB.

Despite the numerous theoretical approaches treating the kinetics of protein binding to nucleic acids, little experimental evidence is available for the kinetics of EcoSSB binding to single-stranded DNA. This is due at least in part to the fact that EcoSSB cannot be dissociated from the standard substrates like natural single-stranded DNAs or poly(dT) by increasing salt concentrations. More data are available for other single-stranded DNA binding proteins (e.g. gene 5 protein of fd-phage (Pörschke and Rauh, 1983) or T4 phage gene 32 protein).

Schneider and Wetmur (1982) were among the first to systematically analyse the dissociation of EcoSSB from single-stranded DNA. Using a ^{3}H-labelled, 375-nucleotides-long denatured pBR322 DNA fragment, complexed with EcoSSB as a donor and single-stranded DNA of different lengths as an acceptor for dissociated EcoSSB, they measured dissociation half-times of the order of 100 s. From this rate, and from the length dependence of the exchange rates, they infer that transfer of EcoSSB proceeds by a direct interaction of the complex with free DNA, rather than by a dissociation–reassociation mechanism.

Similar rates of the order of 0.01 s^{-1} were found for the transfer of EcoSSB from poly(dT) to an excess of (dA)$_{16}$ oligonucleotide (Schomburg, 1985).

Krauss *et al.* (1981) determined the kinetics of the reaction of EcoSSB with single-stranded oligonucleotides of different lengths. They found that the stability of the complex is determined largely by the dissociation rate. d(pT)$_8$, of which there is one molecule bound per monomeric subunit of the EcoSSB tetramer, dissociates with a rate of almost 2000 s^{-1}, whereas the longer oligonucleotides dissociate much more slowly (cf. Table 4.5).

Table 4.5   Intrinsic rate constants for association of EcoSSB to various lengths of oligo-d(pT)

Oligonucleotide	Binding sites	k_R (M^{-1} s^{-1})	k_D (s^{-1})
d(pT)$_8$	4 independent equivalent	$7 \cdot 10^7$	1700
d(pT)$_{16}$	2 independent equivalent	$5 \cdot 10^7$	40
d(pT)$_{30-40}$	1 site	$3 \cdot 10^8$	<1

The apparent rates of 0.01 s^{-1} for dissociating EcoSSB from polymeric single-stranded DNA are in conflict to the *in vivo* situation. The DNA polymerase III can replicate approximately 1000 nucleotides per second, and thus must displace an average of 20 EcoSSB molecules per second (Kornberg, 1982, p. S35). Römer *et al.* (1984) studied the behaviour of EcoSSB molecules bound to poly(dT) by NMR spectroscopy. They found that a single EcoSSB can interact with all methyl groups of (dT)$_{900}$ simultaneously on the NMR time scale. Since it was also known that short oligonucleotides can dissociate fast from a single subunit of EcoSSB the NMR observation was interpreted as a fast translocation of the EcoSSB molecule on the poly(dT), either by a sliding mechanism or by a fast exchange of poly(dT) stretches wrapped round the EcoSSB tetramer.

In determining association rates for polymeric single-stranded DNAs, one has to correctly define the concentration units in which the rate constants are to be expressed. In the following, we define the rate constants on the basis of concentrations of polymers, rather than on nucleotide residues (cf. Lohman, 1986).

The association kinetics of EcoSSB to polymeric single-stranded DNAs has been studied by stopped-flow experiments (Römer *et al.*, 1984; Lohman, 1986; Schomburg, 1985). For a simple bimolecular association,

$$EcoSSB + poly(dT) \rightleftharpoons EcoSSB \cdot poly(dT)$$

at a large excess of binding sites on poly(dT), the reaction proceeds in a pseudo-first-order manner. The rate constants determined by the different authors are given in Table 4.6. All rate constants agree well with the mechanism of a diffusion-limited association (Berg *et al.*, 1981; Lohman, 1986).

Table 4.6   Rate constants for the association of EcoSSB to different polynucleotides[a]

Nucleotide	k_{on} [10^9 M^{-1} s^{-1}]	Added salt	Reference
poly(U)	4.7	<0.1 M NaCl	Lohman, 1986
poly(rεA)	4.5	<0.075 M NaCl	ibid.
poly(dT)	2.4	0.2 M NaCl	Römer *et al.*, 1984
	1.1	0.95 M NaCl	ibid.
poly(dC)	0.9	0.2 M NaCl	Schomburg, 1985
poly(dA)[b]	0.3	0.05 M NaCl	ibid.
	0.11	0.2 M NaCl	ibid.
fd-DNA	0.3	0.2 M NaCl	ibid.

[a] Pseudo-first-order rate constants in excess of binding sites on the polymer over EcoSSB.
[b] In this experiment, the dissociation rate constant could be determined also: 3.3 s^{-1} at 0.05 M NaCl and 7.2 s^{-1} at 0.2 M NaCl.

An interesting observation has been reported by Lohman *et al.* (1986b). At low salt (NaCl<0.1 M), EcoSSB forms a complex with M13mp8 single-stranded DNA in a highly co-operative manner. In these complexes, which can be observed by agarose gel electrophoresis, only fully complexed and completely free M13mp8 single-stranded DNA coexist. These complexes, however, are stable only for about 1 hour; thereafter a redistribution of EcoSSB occurs with a loss of the apparent high co-operativity. The same effect has been observed by us in analytical ultracentrifugation experiments (unpublished). A satisfying explanation for this initially large co-operativity could not be given yet. Lohman *et al.* (1986b) propose that the transient high co-operativity might be a consequence of the slow transition from a complex with 33 nucleotides covered by each EcoSSB tetramer to a complex covering 65 nucleotides (Lohman and Overman, 1985).

A systematic study of the binding of smaller oligonucleotides to EcoSSB has been given by Krauss *et al.* (1981) (see above). For $(dpT)_8$ and $(dpT)_{16}$, the kinetics could be determined by the temperature jump method, whereas for $(dpT)_{30-40}$ the affinity of the oligonucleotide was so high that only the association rate could be observed in a stopped-flow experiment. The results are given in Table 4.5. The association rate constants observed are essentially diffusion-limited, although for the smaller oligonucleotides some pre-equilibrium may exist.

CONCLUDING REMARKS

Many independent studies have shown convincingly that EcoSSB, like other procaryotic SSBs, is needed for a reliable functioning of the many processes involved in replication, recombination, and repair. However, there is little to no information as to the molecular structure of the interactions of EcoSSB. While electron microscopy studies certainly allow some general conclusions with respect to the overall shape of the protein and its complexes with single-stranded DNA, attempts to resolve molecular details, e.g. by X-ray analysis of the protein, have been unsuccessful so far. It can be expected, however, that a combination of site-directed mutagenesis and physicochemical investigations, e.g. spectroscopy, scattering or hydrodynamics, will give further insight into these structural details.

A wealth of data has been obtained describing the interaction of EcoSSB with polymeric and oligomeric nucleic acids. EcoSSB binds exclusively to single-stranded templates. It generally binds DNA more strongly than RNA. It also binds to pyrimidine-homopolynucleotides more strongly than to poly(dA); this difference, however, most probably does not have any functional implications *in vivo*, since all heteropolynucleotides bind with affinities comparable to those of poly(dC) and poly(dT).

Binding of EcoSSB to polymeric lattices is co-operative. After earlier reports of very high co-operativities, it is now well established that the co-operativity parameter is of the order of 100. This co-operativity, together with the high affinity and a number of 500 to 1000 copies of EcoSSB per cell, is sufficient to saturate all single-stranded DNA *in vivo*.

EcoSSB bound to single-stranded DNA can translocate rapidly on its template. This mobility may be essential for the physiological function. EcoSSB acts stoichiometrically. During replication, it has to follow the movement of the replication fork. This can be achieved by dissociating the protein in front of the polymerase and reassociating at the site next to the helicase in form of a treadmilling process as proposed by Hill and Tsuchiya (1981) and taken up by Watson *et al.* (1987). Dissociation rates measured *in vitro*, however, are too slow to be compatible with such a mechanism. Proteins of the replication complex may promote the dissociation of EcoSSB, but so far such a function could not be demonstrated. The problem presented by the slow dissociation is avoided if EcoSSB utilizes its high mobility on the DNA and moves along the template strand simultaneously with the polymerase. The answer to this question has to await a more detailed understanding of replication.

ACKNOWLEDGEMENT

We thank Dr T. M. Lohman for critically reading the manuscript and Dr J. Casas-Finet for providing manuscripts prior to publication.

REFERENCES

Alberts, B. M., Amodio, F. J., Jenkins, M., Gutmann, E. D. and Ferris, R. L. (1968). Studies with DNA-cellulose chromatography. I. DNA-binding proteins from *Escherichia coli*. *Cold Spring Harbour Symp. Quant. Biol.*, **33**, 289–305

Anderson, R. A. and Coleman, J. E. (1975). Physicochemical properties of DNA binding proteins: gene 32 protein of T4 an *E. coli* unwinding protein. *Biochemistry*, **14**, 5485–5491

Babul, J. and Stellwagen, E. (1969). Measurement of protein concentration with interference optics. *Anal. Biochem.*, **28**, 216–221

Baluch, J., Chase J. W. and Sussman, R. (1980). Synthesis of recA protein and induction of bacteriophage lambda in single-strand DNA binding protein mutants of *E. coli. J. Bacteriol.*, **144**, 489–498

Bandyopadhyay, P. K. and Wu, C. (1978). Fluorescence and chemical studies on the interaction of *E. coli* DNA-binding protein with single-stranded DNA. *Biochemistry*, **17**, 4078–4084

Bayer, I., Fliess, A., Greipel, J., Urbanke, C. and Maass, G. (1989). Modulation of the affinity of the single-stranded DNA-binding protein of *Escherichia coli* (*E. coli* SBB) to poly(dT) by site-directed mutagenesis, *Eur. J. Biochem.*, in press

Benham, C. (1985). Theoretical analysis of conformational equilibria in superhelical DNA. *Ann. Rev. Biophys. Biophys. Chem.*, **14**, 23–45

Berg, O. G. and von Hippel, P. (1985). Diffusion controlled macromolecular interactions. *Ann. Rev. Biophys. Biophys. Chem.*, **14**, 131–160

Berg, O.G., Winter, R. B. and von Hippel, P. H. (1981). Diffusion-driven mechanisms of protein translocation on nucleic acids. I. Models and theory. *Biochemistry*, **20**, 6929–6948

Bobst, E. V., Bobst, A. M., Perrino, F. W., Meyer, R. R. and Rein, D. C. (1985). Variability in the nucleic acid binding site size and the amount of single-stranded DNA binding protein in *E. coli. FEBS Letters*, **181**, 133–137

Böhme, H. J., Kopperschläger, G., Schulz, G. and Hofman, E. (1972). Affinity chromatography of phosphofructokinase using Cibacron blue F3G-A. *J. Chrom.*, **69**, 209–214

Boidot-Forget, M., Saeson-Behmoaras, T., Toulme, J.-J. and Helene, C. (1986). Single-strand binding proteins from phage T4 and *E. coli* form nucleosome-like structures with poly(dT). *Biochimie.*, **68**, 1129–1134

Bujalowski, W. and Lohman, T. M. (1986). *E. coli* single-strand binding protein forms multiple, distinct complexes with single-stranded DNA. *Biochemistry*, **25**, 7799–7802

Bujalowski, W. and Lohman, T. M. (1987a). Limited co-operativity in protein–nucleic acid interactions. *J. Mol. Biol.*, **195**, 897–907

Bujalowski, W. and Lohman, T. M. (1987b). A general method of analysis of ligand–macromolecule equilibria using a spectroscopic signal from the ligand to monitor binding. Application to *Escherichia coli* single-strand binding protein–nucleic acid interactions. *Biochemistry*, **26**, 3099–3106

Casas-Finet, J. R., Jhon, N. M., Khamis, M. I., Maki, A. H., Ruvolo, P. P. and Chase, J. W. (1988). An IncY plasmid-encoded single-stranded DNA binding protein from *E. coli* shows identical pattern of stacked Trp residues as the chromosomal ssb gene product. *Eur. J. Biochem.*, **178**, 101–107

Casas-Finet, J. R., Khamis, M. I., Maki, A., Ruvolo, P. R. and Chase, J. W. (1987a). Optically detected magnetic resonance of tryptophan residues in *E. coli* ssb gene product and *E. coli* plasmid-encoded single-stranded DNA-binding proteins and their complexes with poly(deoxythymidylic) acid. *J. Biol. Chem.*, **262**, 8574–8583

Casas-Finet, J. R., Khamis, M. I., Maki, A. and Chase, J. W. (1987b). Tryptophan 54 and phenylalanine 60 are involved synergistically in the binding of *E. coli* SSB protein to single stranded polynucleotides. *FEBS Letters*, **220**, 347–352

Cassuto, E., West, S. C., Mursalim, J., Conlon, S. and Howard-Flanders, P. (1980). Initiation of genetic recombination: Homologous pairing between duplex DNA molecules promoted by recA protein. *Proc. Natl Acad. Sci. USA*, **77**, 3962–3966

Cha, T. A. and Maki, A. H. (1984). Close range interactions between nucleotide bases and Trp residues in an *E. coli* single-strand DNA binding protein–mercurated poly(Uridylic acid) complex. *J. Biol. Chem.*, **259**, 1105–1109

Chase, J. W. (1984). The role of *E. coli* single-stranded DNA binding protein in DNA metabolism. *BioEssays*, **1**, 218–222

Chase, J. M. and Williams, K. R. (1986). Single-stranded DNA binding proteins required for DNA replication. *Ann. Rev. Biochem.*, **55**, 103–136

Chase, J. W., Whittier, R. F., Auerbach, J., Sancar, A. and Rupp, W. D. (1980). Amplification of single-stranded DNA binding protein in *E. coli*. *Nucl. Acids Res.*, **8**, 3215–3227

Chase, J. W., Merrill, B. M. and Williams, K. R. (1983). F sex factor encodes a single-stranded DNA binding protein (SSB) with extensive sequence homology to *Escherichia coli* SSB. *Proc. Natl Acad. Sci. USA*, **80**, 5480–5484

Chase, J. W., L'Italien, J. J., Murphy, J. B., Spicer, E. K. and Williams, K. R. (1984). Characterization of the *E. coli* ssb113 mutant single-stranded DNA binding protein. Cloning of the gene, DNA and protein sequence analysis, HPLC, peptide mapping and DNA binding studies. *J. Biol. Chem.*, **259**, 805–814

Christiansen, C. and Baldwin, R. L. (1977). Catalysis of DNA reassociation by the *Escherichia coli* DNA binding protein. *J. Mol. Biol.*, **115**, 441–454

Chrysogelos, S. and Griffith, J. (1982). *E. coli* single-strand binding protein organizes single-stranded DNA in nucleosome like units. *Proc. Natl Acad. Sci. USA*, **79**, 5803–5807

Chrysogelos, S. and Griffith, J. (1984). Visualization of SSB-ssDNA complexes active in the assembly of stable RecA-DNA filaments. *Cold Spring Harbour Symp. Quant. Biol.*, **49**, 553–559

Clore, G. M. and Gronenborn, A. (1984). An investigation into the solution structure of the single-stranded DNA undecamer 5'-dAAGTGTGATAT by means of nuclear Overhauser enhancement measurements. *Eur. Biophys. J.*, **11**, 95–102

Clore, G. M., Gronenborn, A. M., Greipel, J. and Maass, G. (1986). Conformation of the DNA undecamer 5'd(AAGTGTGATAT) bound to the single-stranded DNA binding protein of *E. coli*. A time-dependent transferred nuclear Overhauser study. *J. Mol. Biol.*, **187**, 119–124

Cohen, S. P., Resnick, J. and Sussman, R. (1983). Interaction of single-strand binding protein and RecA protein at the single-stranded DNA site. *J. Mol. Biol.*, **167**, 901–910

Cox, M. M., Soltis, D. A., Lehman, I. R., De Brosse, C. and Benkobovic, S. J. (1983). On the role of single-stranded DNA binding protein promoted DNA strand exchange. *J. Biol. Chem.*, **258**, 2586–2592

Egner, C., Azhderian, E., Tsang, S. S., Radding, C. M. and Chase, J. W. (1987). Effects of various single-strand DNA binding proteins on reactions promoted by recA protein. *J. Bacteriol.*, **169**, 3422–3428

Epstein, I. R. (1978). Co-operative and non-co-operative binding of large ligands to a finite one-dimensional lattice. A model for ligand–oligonucleotide interactions. *Biophys. Chem.*, **8**, 327–339

Fay, P. J., Johanson, K. O., McHenry, C. S. and Bambara, R. A. (1981). Size classes of products synthesized processively by DNA polymerase III and DNA polymerase III holoenzyme of *E. coli*. *J. Biol. Chem.*, **256**, 976–983

Field, J., Gronostajski, R. M. and Hurwitz, J. (1984). Properties of the adenovirus DNA polymerase. *J. Biol. Chem.*, **259**, 9487–9495

Glikin, G. C., Gargiulo, G., Rena-Descalzi, L. and Worcel, A. (1983). *E. coli* single-stranded DNA binding protein stabilizes specific denatured sites in superhelical DNA. *Nature*, **303**, 770–774

Greipel, J., Maass, G. and Mayer, F. (1987). Complexes of the single-strand DNA binding protein from *E. coli* (EcoSSB) with poly(dT). An investigation of their structure and internal dynamics by means of electron microscopy and nmr. *Biophys. Chem.*, **26**, 149–161

Grosse, F., Nasheuer, H. P., Scholtissek, S. and Schomburg, U. (1986). LDH and GAPDH are single-stranded DNA binding proteins that affect the DNA-polymerase α primase complex. *Eur. J. Biochem.*, **160**, 459–467

Hilgenfeld, R., Saenger, W., Schomburg, U. and Krauss, G. (1984). Novel crystal forms of a proteolytic core of the single-stranded DNA binding protein (SSB) from *E. coli*. *FEBS Letters*, **170**, 143–146

Hill, T. L. and Tsuchiya, T. (1981). Theoretical aspects of translocation on DNA: adenosine triphosphatases and treadmilling binding proteins. *Proc. Natl Acad. Sci. USA*, **78**, 4796–4800

Jong, A. Y. S. and Campbell, J. L. (1984). The CDC-8 gene product of yeast encodes thymidilate kinase. *J. Biol. Chem.*, **259**, 1052–1059

Khamis, M. I., Casas-Finet, J. R., Maki, A. H., Murphy, J. B. and Chase, J. W. (1987a). Role of tryptophan 54 in the binding of *E. coli* single-stranded DNA-binding protein to single-stranded polynucleotides. *FEBS Letters*, **211**, 155–159

Khamis, M. I., Casas-Finet, J. R. and Maki, A. H. (1987b). Stacking interactions of Trp residues and nucleotide bases in complexes formed between *E. coli* single-stranded DNA binding protein and heavy atom modified poly(U). *J. Biol. Chem.*, **262**, 1725–1733

Khamis, M. I., Casas-Finet, J. R., Maki, A. H., Murphy, J. B. and Chase, J. W. (1987c). Investigation of the role of individual tryptophan residues in the binding of *E. coli* single-stranded DNA binding protein to single-stranded polynucleotides. *J. Biol. Chem.*, **262**, 10938–10945

Khamis, M. I., Casas-Finet, J. R., Maki, A. H., Ruvolo, P. P. and Chase, J. W. (1987d). Optically detected magnetic resonance of Trp residues in complexes formed between a bacterial single-stranded DNA binding protein and heavy atom modified poly(U). *Biochemistry*, **26**, 3347–3354

Kornberg, A. (1980). *DNA Replication*. W. H. Freeman & Co., San Francisco

Kornberg, A. (1982). *1982 Supplement to DNA Replication*. W. H. Freeman & Co., San Francisco

Kowalczykowski, S. C., Bear, D. G. and von Hippel, P. H. (1981). Single-stranded DNA binding proteins. *The Enzymes*, vol. 14 (P. Boyer, ed.), Academic Press, New York

Kowalczykowski, S. C., Clow, J., Somani, R. and Varghese, A. (1987). Effects of the *E. coli* SSB protein on the binding of *E. coli* RecA protein to single stranded DNA. Demonstration of competitive binding and the lack of a specific protein–protein interaction. *J. Mol. Biol.*, **191**, 81–95

Kowalczykowski, S. C. and Rupp, R. A. (1987). Effects of *Escherichia coli* SSB protein on the single-stranded DNA-dependent ATPase activity of *Escherichia coli* RecA protein. *J. Mol. Biol.*, **193**, 97–113

Krauss, G., Sindermann, H., Schomburg, U. and Maass, G. (1981). *E. coli* single-stranded deoxyribonucleic acid binding protein: stability, specificity, and kinetics of complexes with oligonucleotides and deoxyribonucleic acid. *Biochemistry*, **20**, 5346–5352

Kunkel, T. A., Meyer, R. R. and Loeb, L. A. (1979). Single-stranded DNA binding protein enhances the fidelity of DNA synthesis in vitro. *Proc. Natl Acad. Sci. USA*, **76**, 6331–6335

Langowski, J., Benight, A. S., Fujimoto, B. S., Schurr, J. M. and Schomburg, U. (1985). Change of conformation and internal dynamics of super-coiled DNA upon binding of *E. coli* single-stranded binding protein. *Biochemistry*, **24**, 4022–4028

Lohman, T. M. (1983). Cooperatively bound protein nucleic acid complexes. *Biopolymers*, **22**, 1697–1713

Lohman, T. M. (1986). Kinetics of protein-nucleic acid interactions: use of salt effects to probe mechanisms of interaction. *CRC Critical Reviews in Biochemistry*, **19**, 191–245

Lohman, T. M. and Kowalczykowski, S. C. (1981). Kinetics and mechanism of the association of the bacteriophage T4 gene 32 (helix destabilizing) protein with single-stranded nucleic acids. *J. Mol. Biol.*, **152**, 67–110

Lohman, T. M. and Overman, L. B. (1985). Two binding modes in *E. coli* single-strand binding protein (SSB) single-stranded DNA complexes: modulation by NaCl concentration. *J. Biol. Chem.*, **260**, 3594–3603

Lohman, T. M., Green, J. M. and Beyer, R. S. (1986a). Large-scale overproduction and rapid purification of the *E. coli* ssb gene product. Expression of the ssb gene under λ P_L control. *Biochemistry*, **25**, 21–25

Lohman, T. M., Overman, L. B. and Datta, S. (1986b). Salt-dependent changes in the DNA binding co-operativity of *E. coli* single-stranded binding protein. *J. Mol. Biol.*, **187**, 603–616

McEntee, K. (1985). Kinetics of DNA renaturation catalyzed by recA protein of *E. coli*. *Biochemistry*, **24**, 4345–4351

McEntee, K., Weinstock, G. M. and Lehman, I. R. (1980). RecA protein-catalyzed strand assimilation: stimulation by *E. coli* single-stranded DNA binding protein. *Proc. Natl Acad. Sci. USA*, **77**, 857–861

McGhee, J. D. and von Hippel, P. H. (1974). Theoretical aspects of DNA-protein interactions: co-operative and non-co-operative binding of large ligands to a one-dimensional homogeneous lattice. *J. Mol. Biol.*, **86**, 469–489

Mazur, S. J. and Record, M. T. (1986). Kinetics of nonspecific binding reactions of proteins with DNA flexible coils: site-based and molecule-based association reactions. *Biopolymers*, **25**, 985–1008

Merrill, B. M., Williams, K. R., Chase, J. W. and Konigsberg, W. H. (1984). Photochemical cross-linking of the *E. coli* single-stranded DNA binding protein to oligodeoxynucleotides. *J. Biol. Chem.*, **259**, 10850–10856

Meyer, R. R., Glassberg, J. and Kornberg, A. (1979). An *E. coli* mutant defective in single-strand binding protein is defective in DNA replication. *Proc. Natl Acad. Sci. USA*, **76**, 1702–1705

Meyer, R. R., Glassberg, J., Scott, J. V. and Kornberg, A. (1980). Temperature-sensitive single-stranded DNA-binding protein from *Escherichia coli*. *J. Biol. Chem.*, **255**, 2897–2901

Molineux, I. J., Friedman, S. and Gefter, M. L. (1974). Purification and properties of the *E. coli* deoxyribonucleic acid-unwinding protein. *J. Biol. Chem.*, **249**, 6090–6098

Molineux, I. J. and Gefter, M. (1974). Properties of the *Escherichia coli* DNA binding (unwinding) protein: Interaction with DNA polymerase and DNA. *Proc. Natl Acad. Sci. USA*, **71**, 3858–3862

Molineux, I. J. and Gefter, M. (1975). Properties of the *Escherichia coli* DNA binding (unwinding) protein: interaction with nucleolytic enzymes and DNA. *J. Mol. Biol.*, **98**, 811–825

Molineux, I. J., Pauli, A. and Gefter, M. L. (1975). Physical studies of the interaction between the *Escherichia coli* DNA binding protein and nucleic acids. *Nucl. Acids Res.*, **2**, 1821–1837

Monzingo, A. F. and Christiansen, C. (1983). Crystallization of single-stranded DNA binding protein. *J. Mol. Biol.*, **170**, 797–801

Morrical, S. W., Lee, J. and Cox, M. M. (1986). Continuous association of *E. coli* single-strand DNA binding protein with stable complexes of recA protein and single-stranded DNA. *Biochemistry*, **25**, 1482–1494

Niyogi, S. K., Ratrie, H., III and Datta, A. K. (1977). Effect of *E. coli* DNA binding protein on the transcription of single-stranded phage M13 DNA by *E. coli* RNA polymerase. *Biochem. Biophys. Res. Comm.*, **78**, 343–349

Ollis, D., Brick, P., Abdel-Meguid, S. S., Murthy, K., Chase, J. W. and Steitz, T. A. (1983). Crystals of *E. coli* single-strand DNA binding protein show that the tetramer has D_2 symmetry. *J. Mol. Biol.*, **170**, 797–800

Overman, L. B., Bujalowski, W. and Lohman, T. M. (1988). Equilibrium binding of *E. coli* single-strand binding protein to single-stranded nucleic acids in the $(SSB)_{65}$ binding mode. Cation and anion effects and polynucleotide specificity. *Biochemistry*, **27**, 465–471

Perrino, F. W., Rein, D. C., Ruben, S. M., Bobst, A. M. and Meyer, R. R. (1986). Protein–protein interactions of *E. coli* single-stranded DNA binding protein identified by SSB affinity chromatography. UCLA Symposia on Molecular and Cellular Biology, Mar. 16–23, *J. Cell Biochem. Suppl.*, 239

Perucho, M., Salas, J. and Salas, M. L. (1977). Identification of mammalian DNA binding protein P8 as glyceraldehyde-3-phosphate dehydrogenase. *Eur. J. Biochem.*, **81**, 557–562

Pörschke, D. and Rauh, H. (1983). Cooperative, excluded site binding and its dynamics for the interaction of gene 5 protein with polynucleotides. *Biochemistry*, **22**, 4737–4745

Radding, C. M., Flory, J., Wu, A., Kahn, R., DasGupta, C., Gonda, D., Bianchi, M. and Tsang, S. S. (1982). Three phases in homologous pairing. Polymerization of recA protein on single-stranded DNA, synapsis and polar strand exchange. *Cold Spring Harbour Symp. Quant. Biol.*, **47**, 821–828

Reckmann, B., Grosse, F., Urbanke, C., Frank, R. and Blöcker, H. (1985). Analysis of secondary structures in M13mp8(+) single-stranded DNA by the pausing of DNA polymerase α. *Eur. J. Biochem.*, **152**, 633–644

Register, J. C. II and Griffith, J. (1985). The direction of RecA protein assembly onto single-stranded DNA is the same as the direction of strand assimilation during strand exchange. *J. Biol. Chem.*, **260**, 12308–12312

Resnick, J. and Sussman, R. (1982). *E. coli* single-strand DNA binding protein from wild type and lexC113 mutant affects in vitro proteolytic cleavage of phage λ repressor. *Proc. Natl Acad. Sci. USA*, **79**, 2832–2835

Richter, A., Sapp, M. and Knippers, R. (1986). Are single-strand specific DNA binding proteins needed for mammalian replication? *Trends Biochem. Sci.*, **11**, 283–284

Riddles, P. W. and Lehman, I. R. (1985a). The formation of paranemic and plectonemic joints between DNA molecules by the recA and single-stranded DNA-binding proteins of *Escherichia coli*. *J. Biol. Chem.*, **260**, 165–169

Riddles, P. W. and Lehman, I. R. (1985b). The formation of plectonemic joints by the recA protein of *Escherichia coli*. *J. Biol. Chem.*, **260**, 170–173

Römer, R., Schomburg, U., Krauss, G. and Maass, G. (1984). *E. coli* single-stranded DNA binding protein is mobile on DNA. ^{1}H NMR study of its interaction with oligo- and polynucleotides. *Biochemistry*, **23**, 6132–6137

Ruyechan, W. T. and Wetmur, J. G. (1975). Studies on the co-operative binding of the *E. coli* DNA unwinding protein to single-stranded DNA. *Biochemistry*, **14**, 5529–5534

Ruyechan, W. T. and Wetmur, J. G. (1976). Studies on the non-co-operative binding of the *E. coli* DNA unwinding protein to single-stranded nucleic acids. *Biochemistry*, **15**, 5057–5062

Sancar, A., Williams, K. R., Chase, J. W. and Rupp, W. D. (1981). Sequences of the ssb gene and protein. *Proc. Natl Acad. Sci. USA*, **78**, 4274–4278

Schneider, R. J. and Wetmur, J. G. (1982). Kinetics of transfer of *E. coli* single-stranded DNA binding protein between single-stranded DNA molecules. *Biochemistry*, **21**, 608–615

Scholtissek, S. and Grosse, F. (1988). A plasmid vector system for the expression of a tri-protein consisting of β-galactosidase, a collagenase recognition site and a foreign gene. *Gene*, **62**, 55–64

Schomburg, U. (1985). *Escherichia coli* single-strand DNA binding protein: investigations on the stability, kinetics, and structure of its complexes with nucleic acids. Thesis, University of Hanover, West Germany

Schwarz, G. and Watanabe, F. (1983). Thermodynamics and kinetics of co-operative protein–nucleic acid binding. I. General aspects of analysis of data. *J. Mol. Biol.*, **163**, 467–484

Schwarz, G. and Stankowski, S. (1979). Linear co-operative binding of large ligands involving mutual exclusion of different binding modes. *Biophys. Chem.*, **10**, 173–181

Schwarz, G. (1977). On the analysis of linear binding effects associated with curved scatchard plots. *Biophys. Chem.*, **6**, 65–76

Shimamoto, N., Ikushima, N., Utiyama, H., Tachibana, H. and Horie, K. (1987). Specific and co-operative binding of *E. coli* single-strand DNA binding protein to mRNA. *Nucl. Acids Res.*, **15**, 5241–5250

Sigal, N., Delius, J., Kornberg, T., Gefter, M. L. and Alberts, B. (1972). A DNA-unwinding protein isolated from *Escherichia coli*: its interaction with DNA and with DNA polymerases. *Proc. Natl Acad. Sci. USA*, **69**, 3537–3541

Srivenugopal, K. S. and Morris, D. R. (1986). Modulation of the relaxing activity of *E. coli* topo-isomerase I by single-stranded DNA binding proteins. *Biochem. Biophys. Res. Commun.*, **137**, 795–800

Valentini, O., Biamonti, G., Pandolfo, M., Morandi, C. and Riva, S. (1985). Mammalian single-stranded DNA binding proteins and heterogeneous nuclear RNA proteins have common antigenic determinants. *Nucl. Acids Res.*, **13**, 337–346

VanAmerongen, H., VanGrondelle, R. and VanDeVliet, P. C. (1987). Interaction between adenovirus DNA binding protein and single-stranded polynucleotides studied by CD and UV absorption. *Biochemistry*, **26**, 4646–4652

VanDerEnde, A., Baker, T. A., Ogawa, T. and Kornberg, A. (1985). Initiation of enzymatic replication at the origin of the *E. coli* chromosome: primase as the sole priming enzyme. *Proc. Natl Acad. Sci. USA*, **82**, 3954–3958

VanMansfeld, A. D. M., VanTeeffelen, H. A. A. M., Fluit, A. C., Baas, P. D. and Jansz, H. S. (1986). Effect of SSB protein on cleavage of single-stranded DNA by ΦX gene A protein and A* protein. *Nucl. Acids Res.*, **14**, 1845–1861

Walker, G. C. (1984). Mutagenesis and inducible responses to deoxyribonucleic acid damage in *Escherichia coli. Microbiological Reviews*, Mar. 1984, 60–93

Watson, J. D., Hopkins, N. H., Roberts, J. W., Steitz, J. and Weiner, A. M. (1987). *Molecular Biology of the Gene*, vol. **1**, Benjamin/Cummings, Menlo Park, 292

Weiner, J. H., Bertsch, L. L. and Kornberg, A. (1975). The deoxyribonucleic acid unwinding protein of *Escherichia coli. J. Biol. Chem.*, **250**, 1972–1980

West, S. C., Cassuto, E. and Howard-Flanders, P. (1982). Role of SSB protein in recA promoted branch migration reactions. *Mol. Gen. Genet.*, **186**, 333–338

Williams, R. C. and Spengler, S. J. (1986). Fibers of RecA protein and complexes of RecA protein and single-stranded ΦX 174 DNA as visualized by negative-stain electron microscopy. *J. Mol. Biol.*, **187**, 109–118

Williams, K. R., Spicer, E. K., Lopresti, M. B., Guggenheimer, R. A. and Chase, J. W. (1983). Limited proteolysis studies on the *E. coli* single-stranded DNA binding protein. *J. Biol. Chem.*, **258**, 3346–3355

Williams, K. R., Murphy, J. B. and Chase, J. W. (1984). Characterization of the structural and functional defect in the *E. coli* single-stranded DNA binding protein encoded by the ssb1 gene. *J. Biol. Chem.*, **259**, 11804–11811

Williams, K. R., Reddigari, S. and Patel, G. L. (1985). Identification of a nucleic acid helix destabilizing protein from rat liver as LDH 5. *Proc. Natl Acad. Sci. USA*, **82**, 5260–5264

Woodbury, C. P., Jr (1981). Free sliding ligands – an alternative model of DNA-protein interactions. *Biopolymers*, **20**, 2225–2241

Zang, L. H., Maki, A. H., Murphy, J. B. and Chase, J. W. (1987). Triplet state sublevel kinetics of Trp 54 in the complex of *E. coli* single-strand DNA binding protein with single-stranded poly(dT). *Biophys. J.*, **52**, 867–873

5

Protein–nucleic acid interactions in tobacco mosaic virus

Gerald Stubbs

INTRODUCTION

Tobacco mosaic virus (TMV), a positive-strand RNA virus, provides a unique perspective on nucleic acid structure, in that it has been for some years the only macromolecular assembly in which the structure of a ribonucleic acid interacting with a protein has been visualized in molecular detail, and one of the very few showing protein–nucleic acid interactions at all. Crystallographers have determined the structures of a number of spherical RNA viruses (for references see Liljas, 1986; Stubbs, 1989); but in those structures so far determined, the nucleic acid is not seen in the electron density maps, because it does not conform to the icosahedral symmetry of the coat proteins. TMV, by contrast, is a helical virus, and its RNA is well ordered, with the same symmetry as the protein. It is the type member of the tobamovirus group, rod-shaped viruses 3000 Å long and 180 Å in diameter, with a central hole of diameter 40 Å. Approximately 2130 identical protein subunits of molecular weight 17500 form a right-handed helix of pitch 23Å with 49 subunits in 3 turns. A single strand of RNA follows the basic helix between the protein subunits at a radius of 40 Å, with three nucleotides bound to each protein subunit. An overall view of part of the viral rod is shown in Figure 5.1.

   The RNA of TMV was first isolated by Fraenkel-Conrat and Williams (1955), who were able to dissociate the virus in SDS, sufficiently free of degradation to be able to reconstitute infectious virus from the protein and RNA. The reconstitution was a milestone in itself, establishing TMV as a paradigm of self-assembly, a system that goes through a highly structured sequence of assembly events without the aid of external molecular machinery, or temporary structures such as scaffolding proteins. The first step in the assembly sequence is the binding of the RNA to a 20S aggregate of the protein (Butler and Klug, 1971; Ohno et al., 1972), and Zimmern

Figure 5.1   Computer graphics representation of about one-twentieth of the TMV particle. Protein subunits are light grey; RNA is dark grey. The RNA is shown extending beyond the end of the protein helix for clarity. There are 49 protein subunits in 3 turns of the viral helix, and 3 nucleotides bound to each protein subunit. Graphics from Namba *et al.* (1985)

and Wilson (1976) showed that this binding is initiated at a site about one-sixth of the way along the nucleic acid. Zimmern (1977) determined the sequence of this site, and proposed the RNA secondary structure shown in Figure 5.2. The entire nucleotide sequence was determined by Goelet *et al.* (1982).

The 20S aggregate was originally considered to be a disk containing two layers of 17 subunits each. The disk has been crystallized, and, despite its large size (molecular weight about 600 000), its structure has been determined at 2.8 Å resolution, making use of the 17-fold symmetry in the structure determination as well as the more usual methods of isomorphous replacement (Bloomer *et al.*, 1978). The atomic coordinates of the protein in the disk have been refined (Mondragon, 1984). The aggregate binds to the initiation site in such a way that the 3' tail of the RNA is on one side of

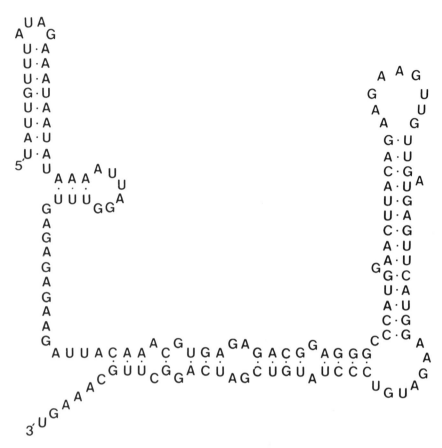

Figure 5.2 The origin of assembly sequence in TMV RNA, folded into the secondary structure proposed by Zimmern (1977). Note the AAGAAGUUG sequence in the hairpin loop

the protein aggregate, and the much longer 5' tail is looped back through the hole in the centre of the aggregate. The assembly process (Figure 5.3) continues by adding protein to the 5' end, pulling the RNA through the hole and co-operatively binding nucleotides to the growing nucleoprotein complex (Butler *et al.*, 1977; Lebeurier *et al.*, 1977). Recent results (Correia *et al.*, 1985; Namba and Stubbs, 1986), suggest that the 20S aggregate may actually be a short helix (protohelix) of about 2½ turns, containing about 39 subunits, but this interpretation would not alter the essential description of the process of initiation of viral assembly.

X-ray fibre diffraction patterns from oriented gels of TMV were recorded by Bernal and Fankuchen (1941). The structure was determined at 4 Å resolution, using multi-dimensional isomorphous replacement applied to fibre diffraction data (Stubbs *et al.*, 1977); the RNA conformation

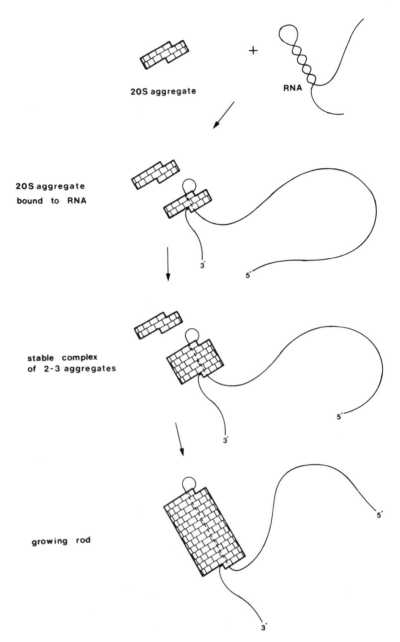

20S aggregate

+

RNA

20S aggregate
bound to RNA

3'

5'

stable complex
of 2-3 aggregates

5'

3'

growing rod

5'

3'

Figure 5.3 Early stages in the assembly of TMV. Some authors consider the nucleating aggregate to be the 34-subunit disk, rather than the protohelix

was described by Stubbs and Stauffacher (1981). Multi-dimensional iso-morphous replacement is more akin to protein crystallography than it is to traditional methods of fibre diffraction, depending on objective phase information from heavy-atom derivatives rather than models with their inherent assumptions. The resolution of the 1977 analysis deteriorated near the outer surface of the virus because of technical problems with fibre diffraction at that time, but these problems were overcome in the 3.6 Å analysis of Namba and Stubbs (1985; 1986). Resolution has now been extended to 2.9 Å (Namba *et al.*, 1989); the refined 2.9 Å structure permits a detailed description of the RNA, the coat protein and their interactions.

NUCLEOTIDE SEQUENCE

Goelet *et al.* (1982) determined the complete sequence of the viral RNA for TMV *vulgare*. The sequence of a tomato strain (L) has been reported by Ohno *et al.* (1984), and partial sequences, in particular of the 3' ends of the RNA, have been published for several other strains, including the cowpea strain, also called sunn-hemp mosaic virus (Meshi *et al.*, 1981), the Japanese common strain OM (Meshi *et al.*, 1982) and cucumber green mottle mosaic virus, watermelon strain, CGMMV-W (Meshi *et al.*, 1983).

The RNA is capped at the 5' end by the sequence $m^7G^{5'}ppp^{5'}G \cdots$ (Zimmern, 1975; Keith and Fraenkel-Conrat, 1975), followed by a tract of 69 nucleotides without any G residues, terminated by the first UAG initiation codon (Richards *et al.*, 1978). The cap is necessary for infectivity, but not for viral assembly (Ohno *et al.*, 1976), and the cap together with the untranslated leader sequence have been shown to enhance expression of mRNA (Gallie *et al.*, 1987a, b). A number of subgenomic messenger RNAs are also capped (Guilley *et al.*, 1979; Fukuda *et al.*, 1980).

There are many minor instances of polymorphism throughout the sequence, but the only major case is a significant dimorphism at the 5' end of *vulgare*. Ohno *et al.* (1984) have questioned the existence of this dimorphism, showing that one of the two sequences is almost identical to that of the tomato strain, but Bloomer and Butler (1986) point out that the method of sequencing used by Goelet *et al.* (1982) was extremely unlikely to have allowed a contamination of the virus by the tomato strain to have gone unnoticed.

TMV RNA codes for at least four proteins, in three open reading frames (Figure 5.4). Other possible proteins are discussed by Goelet *et al.* (1982). The first frame begins at residue 69, and codes for a protein of molecular weight 126 000, as well as for a protein having the same N-terminus but molecular weight 183 000, synthesized by reading through the amber termination codon. Homologies have been found between these proteins and non-structural proteins from alfalfa mosaic virus, brome mosaic virus and Sindbis virus (Ahlquist *et al.*, 1985), although the structural proteins of the viruses are quite different. Other proteins coded by TMV RNA are derived from subgenomic messenger RNAs, all of which have the same 3′ terminus as the genomic RNA (Figure 5.4). The 5′ ends of these mRNAs have been noted to be rich in U and A, as are the 5′ ends of the genomic RNA and many other plant viral RNAs (Goelet *et al.*, 1982). The first subgenomic mRNA, I_1 (Sulzinski *et al.*, 1985), is believed to direct the synthesis of a protein of molecular weight about 54 000, although unlike the other proteins mentioned here, such a protein has not yet been observed *in vivo*. This protein would be the C-terminal part of the 183K protein, complementary to the 126K protein. The second subgenomic mRNA, I_2 (Beachy and Zaitlin, 1977), directs the synthesis of a protein of molecular weight 30 000, whose cistron overlaps that of the 183K protein by five codons in both *vulgare* and tomato mosaic virus. A third mRNA, LMC-RNA, begins before the end of the cistron for the 30K protein, and contains the third open reading frame, coding for the coat protein (Hunter *et al.*, 1976; Siegel *et al.*, 1976). In *vulgare* and tomato mosaic virus, there are two nucleotides between the 30K protein cistron and the coat protein cistron, but in sunn-hemp virus and CGMMV-W, the cistrons overlap by 26 nucleotides.

The site of initiation of assembly (Figure 5.2) is located about 900 nucleotides from the 3′ end of the RNA in TMV, toward the C-terminal end of the 30K protein cistron. It is characterized by the occurrence of G at every third residue in six consecutive positions, and in equivalent positions in many neighbouring triplets. The sequence also has a low proportion of C residues. The origin of assembly site is found in the same location in tomato mosaic virus and the OM strain of TMV. A sequence of 27 nucleotides is perfectly conserved in tomato mosaic virus, although the homology between the two viruses in this cistron as a whole is only 70 per cent (Ohno *et al.*, 1984). In sunn-hemp virus and CGMMV-W, however, the origin of assembly is in the coat protein cistron, only 400 nucleotides from the 3′ end (Fukuda *et al.*, 1980; Meshi *et al.*, 1981; Meshi *et al.*, 1983). The repeating pattern of Gs is not as pronounced in these two viruses, but there is still an extremely strong tendency for purines to occupy every third site. Because of the location of the origin of assembly in the viruses of this subgroup, these viruses are characterized by the presence of short, defective particles in which only the subgenomic mRNAs have been encapsidated (Bruening *et al.*, 1976; Higgins *et al.*, 1976; Fukuda *et al.*, 1981).

After digestion of TMV RNA with ribonuclease T_1, Guilley *et al.* (1975a, b) found a nucleotide sequence in the coat protein cistron of TMV, about 40 nucleotides from the sunn-hemp and CGMMV-W origin of assembly (Meshi *et al.*, 1983), that binds specifically to the 20S coat protein aggregate. They called this sequence SERF (specifically encapsidated RNA fragment). It has a pronounced tendency for purines to appear every

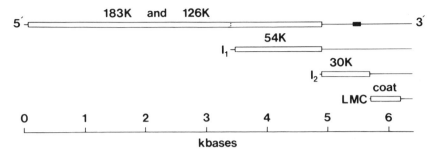

Figure 5.4   The open reading frames (open boxes) of the TMV genome. The 183K and 126K proteins are synthesized using the genomic RNA as a messenger; the 54K, 30K, and coat (17.5K) proteins are derived from the subgenomic messenger RNAs I_1, I_2, and LMC, respectively. The solid box around residue 5478 is the origin of assembly. Data from Goelet *et al.* (1982) and Sulzinski *et al.* (1985)

three residues, and it can be folded into a stable hairpin structure similar to that of the true assembly nucleation site (Hirth and Richards, 1981). In intact TMV RNA, this fragment does not act as an alternative nucleation site, suggesting that the RNA specificity of TMV protein is a function of RNA tertiary structure as well as sequence. This is further demonstrated by the fact that removal of a few nucleotides from the 5′ end of TMV RNA, far away from the nucleation site, prevents assembly, whereas further removal of 1000–5000 nucleotides restores assembly competence (Wilson *et al.*, 1978). Meshi *et al.* (1981) have noted extensive homology between a sequence in sunn-hemp virus RNA about 870 residues from the 3′ end and the origin of assembly in TMV RNA, but unlike the true origin of assembly sequences and the SERF fragment, this sequence cannot be folded into a highly base-paired loop.

The RNA contains no sequence repeats that might reflect interactions between the protein and the nucleic acid other than those in the origin of assembly sequences (Goelet *et al.*, 1982; Butler, 1984). There is a frequent occurrence of the sequence GXX within each reading frame, but the reading frames are not in phase.

Like many other plant virus RNAs, the 3′ ends of the tobamoviruses can be aminoacylated, usually with histidine (Öberg and Philipson, 1972; Carriquiry and Litvak, 1974). The one exception is sunn-hemp virus, which accepts valine (Beachy *et al.*, 1976). Rietveld *et al.* (1984) proposed a base-pairing scheme, supported by the accessibility of nucleotides to various nucleases, that could fold into a tRNA-like structure. Meshi *et al.* (1981) noted a homology of almost 60 per cent between the 3′ ends of sunn-hemp virus RNA and the RNA of turnip yellow mosaic virus (TYMV), a spherical virus which also accepts valine; they and Rietveld *et al.* (1984) have proposed base-paired foldings that could be applied to both sunn-hemp and TYMV.

SPECIFIC AFFINITIES BETWEEN TMV COAT PROTEIN AND RNA

TMV protein assembles efficiently only with its own or closely related RNA. Within the tobamovirus group, reconstitution between RNA and protein of different strains is often very efficient (Atabekov *et al.*, 1970; Okada *et al.*, 1970), although the yield can be rather low when distantly related members such as ribgrass mosaic and *vulgare* (Holoubeck, 1962) are used. However, whereas many spherical viruses can package a wide variety of viral and non-viral RNAs (reviewed by Dodds and Hamilton, 1976) and even such non-biological polyanions as polyvinyl sulphate (Frist, 1968), TMV reconstitutes only poorly with unrelated viral RNA (see, for example, Atabekov *et al.*, 1970; Fritsch *et al.*, 1973). A very small fraction of non-viral RNA has been observed to assemble with TMV coat protein *in vivo* (Rochon and Siegel, 1984). Assembly is possible with poly-A, poly-I and poly-U, but not with poly-C (Fraenkel-Conrat and Singer, 1964; Lanina *et al.*, 1976). The assembly mechanism with poly-A (and, presumably, with other homopolymers) is not the same as that with viral RNA, however, since it shows second-order dependence on protein concentration (Butler, 1972), rather than the first-order dependence shown by TMV RNA (Butler, 1974).

Small oligonucleotides show considerable dependence of protein-binding capacity on nucleotide sequence. Under viral assembly conditions (pH 7.0, ionic strength 0.1, 20 °C), the trinucleoside diphosphates AAG, CAG, UAG and GAA bind weakly, with binding constants less than 300 M^{-1} (Steckert and Schuster, 1982; confirmed for AAG by Turner *et al.*, 1986). Steckert and Schuster determined the protein-binding constants for 25 trinucleoside diphosphates at pHs close to 5.5, and found relatively strong binding of AAG (binding constant $12 \times 10^3 \text{ M}^{-1}$), with GAA, GAG, AUG and CAG also binding strongly. Assuming that all trimers bind to the protein in the same phase (that is, with the first nucleotide always occupying a particular one of the three nucleotide binding sites on the protein), they concluded that the nucleotide binding site occupied by the base at the 3′ end of the trimer (the third base) is sensitive to base size, favouring purines, whereas the one at the 5′ end is sensitive to both base size and substituents, again favouring purines, but also favouring NH_2 substituents away from the ribose ring (that is, adenine and cytosine). The detailed interpretation of these results does depend somewhat upon whether the 20S nucleating aggregate is a helix, like the protein aggregate at pH 5.5, or a disk, like the crystalline form of the protein (Bloomer *et al.*, 1978); nevertheless, the results are consistent with both the sequence of the origin of assembly and the specific protein–nucleic acid interactions found in the structure of the intact virus (Namba *et al.*, 1989; see below).

The reasons for the weak binding of trimers at pH 7 are not entirely clear. Steckert and Schuster (1982) considered the effect to be electrostatic, with the less negatively charged aggregate at pH 5.5 allowing increased binding of the nucleotides. Turner *et al.* (1986) suggested that even at pH 7.0, AAG binds in a different mode from that of TMV RNA; however, the difference in binding strengths of these oligomers can be largely accounted for simply by considering the size of the oligomer (Steckert and Schuster, 1982). Bloomer and Butler (1986) consider the binding protein aggregate at pH 7.0 to be the disk, while the protein at pH 5.5 is in the helical form, isomorphous with the intact virus (Mandelkow *et al.*, 1981). Even if the aggregate at pH 7.0 is a protohelix (Correia *et al.*, 1985; Namba and Stubbs, 1986), it is not isomorphous with the long low pH helix; NMR studies (Jardetzky *et al.*, 1978) show that the inner peptide loop, which includes part of the trinucleotide binding site, is disordered in the nucleating aggregate, whereas it is ordered in the protein helix.

Because of the occurrence of these sequences in the specific binding loop in the origin of assembly sequence (Figure 5.2), Turner *et al.* (1986) studied the binding of AAGAAG and AAGAAGUUG to TMV coat protein under assembly conditions. They found both these oligonucleotides to bind strongly, with binding constants of about 10^6 M^{-1}. AAGAAG saturated the protein binding sites, but AAGAAGUUG saturated only about 60 per cent of the sites. Despite the strong binding in solution, AAGAAG bound to the protein disk crystals much less strongly, with a binding constant of less than 10^3 M^{-1}. The crystal structure does not hinder access of nucleotides to the binding sites, but it could prevent a change of conformation on nucleotide binding (Bloomer and Butler, 1986). Such an interpretation is supported by the observation that the crystals crack when exposed to high concentrations of the larger oligomers (Graham and Butler, 1979; Bloomer *et al.*, 1981; Turner *et al.*, 1986). These observations would also, of course, be consistent with the possibility that, in solution, the oligomers bind to a helical protein form, which would have a much greater affinity than the disk for nucleotides.

TMV coat protein does not bind DNA. Even single-stranded DNA containing the origin of assembly cannot be encapsidated (Gallie *et al.*, 1987c). Turner *et al.* (1986) found no significant binding of d(AAGAAG) by the protein.

THREE-DIMENSIONAL STRUCTURE OF TMV RNA

Nucleic acid structure

From the refined 2.9 Å resolution structure of TMV (Namba *et al.*, 1989), it is evident that the nucleic acid structure is strongly influenced by its interactions with the protein. This leads to some unusual RNA conformations, a fact that should be considered when modelling structures in which, by contrast with TMV, only the protein and not the nucleic acid structure is known.

The RNA structure is shown in Figure 5.5; the RNA torsion angles are given in Table 5.1. The structure is very similar to that found earlier from a 4 Å resolution map (Stubbs and Stauffacher, 1981), with an r.m.s. difference between the models of less than 1 Å. The conformations of all three ribose rings belong to the C3'-endo group: specifically, riboses 1 and 3 are C2'-exo (a conformation extremely close to C3'-endo and often not distinguished from it), and ribose 2 is C3'-endo. The pseudorotation angles (Altona and Sundaralingam, 1972) for the furanose rings 1, 2 and 3 are, respectively, -21 °, 19 ° and -10 °. Nucleotides 1 and 2 are in the *anti* conformation; when base 3 is a purine, it is *syn*. The conformation of the pyrimidines in position 3 is not known, but there is no reason to suppose that it deviates from the more common *anti*. The *syn* conformation is unusual, but it has been seen in small molecules and in Z-DNA (for references, see Saenger, 1984). The energy difference between the two forms is quite small; crystal packing forces and water of crystallization can stabilize the normally *anti* 4-thiouridine in the *syn* form (Lesyng and Saenger, 1981), and binding to a protein converts *anti* 8-bromoadenosine to *syn* (Abdallah *et al.*, 1975). In TMV, the *syn* conformation appears to be stabilized by base stacking between bases 3 and 1, which are about 3.5 Å apart (Figure 5.5), and, when the base is adenine, it is also stabilized by a hydrogen bond between N6 and the main-chain carbonyl of Thr 89. The combination of a *syn* base with a 3'-endo ribose is particularly rare, but one of the few cases where it has been observed is in another example of a nucleotide structure perturbed by protein binding: the adenine in NAD bound to two of the four subunits of lobster D-glyceraldehyde-3-phosphate dehydrogenase (Moras *et al.*, 1975). The combination also occurs in Z-DNA (Wang *et al.*, 1979).

Table 5.1   RNA dihedral angles (°)

Nucleotide	ζ O3'-P	α P-O5'	β O5'-C5'	γ C5'-C4'	δ C4'-C3'	ε C3'-O3'	χ Sugar-base
1	258	142	206	181	91	154	209
2	109	263	216	348	69	73	186
3	206	143	207	239	88	253	25

Nomenclature for atoms and dihedral angles is according to the IUPAC–IUB Joint Commission on Biochemical Nomenclature (1983) (Saenger, 1984).

Only one of the torsional angles is outside the range previously found in nucleotides, repeating helical nucleic acids or tRNA structures (Saenger, 1984): ε for nucleotide 2 is 72 °, reflecting the fact that phosphate 3 is close to ribose 2, rather than being rotated away. Several theoretical studies (references in Saenger, 1984) have considered such a value of ε to be forbidden, but those studies were based on assumptions about the rest of the nucleotide structure drawn from limited numbers of crystal structures. There are no close contacts or distorted bond lengths or angles as a result of this torsional angle, and the structure is stabilized by a hydrogen bond, about 2.7 Å long, between the ribose 2 O2' and one of the phosphate 3 oxygens. Bases 2 and 3 have γ more typical of 2' than 3' riboses, but energy calculations show that the preference is very slight (Saran *et al.*, 1973), and could easily be overcome by protein binding energy.

The RNA binding site is generally rather crowded; phosphates 2 and 3 are only 4.8 Å apart. The crowding is to some extent imposed by the helical symmetry of the virus; for the RNA to bind at a radius of 40 Å, the length of the repeating trinucleotide can only be 15.4 Å.

Protein–nucleic acid interactions

These may be considered in several classes: electrostatic interactions between the phosphate groups and protein side-chains, non-specific (hydrophobic) interactions between the bases and the protein, and base-specific hydrogen bonds with the protein. Surprisingly, there do not appear to be direct hydrogen bonds between the ribose hydroxyl groups and the protein. These hydroxyl groups are, nevertheless, important to the structure, and are discussed below in the context of nucleic acid (RNA rather than DNA) specificity.

The phosphate groups are, in general, neutralized by arginines from the protein subunit below them, but not always in the form of simple ion pairs. Phosphate 1 and phosphate 2 are close (about 3 Å) to Arg 90 and 92, respectively, but there are also several medium-range (5–7 Å) interactions, between phosphate 1 and Arg 92, phosphate 1 and Arg 41, phosphate 3 and Arg 90, and phosphate 3 and Arg 92. Unexpectedly, there is also an unusually close approach (less than 3.5 Å) between the side chain carboxyl group of Asp 116 and phosphate 2. Electron density between these two groups has been interpreted as a calcium ion (Namba *et al.*, 1989); the calcium ion also utilizes water molecules, the ribose hydroxyl group of nucleotide 1, and the ring oxygen of ribose 3 as ligands. Phosphate–carboxylate calcium-binding sites are not common (Einspahr and Bugg, 1984), but one has been observed in staphylococcal nuclease (Cotton *et al.*, 1979), and the metal binding site of DNA polymerase I utilizes three carboxylate groups and a nucleotide phosphate (Ollis *et al.*, 1985). The interaction of the phosphate, carboxylate and calcium in TMV, illustrated in Figure 5.6, is presumed to play an important part in the assembly and disassembly of the virus, as discussed below.

Non-base-specific protein–nucleic acid interactions are essential to ensure that the coat protein can encapsidate the entire RNA genome. All three bases lie flat against one of the protein α-helices, termed (Champness

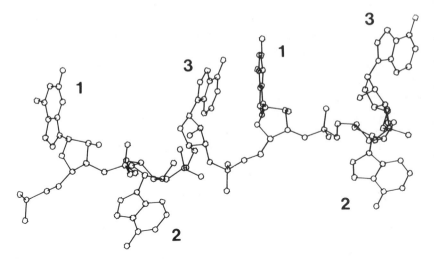

Figure 5.5 Two consecutive trinucleotides of TMV RNA drawn as GAAGAA, viewed approximately from the outside of the virus (compare with Figure 5.7 to see the orientation relative to the protein). Nucleotides are labelled 1, 2 and 3 for reference in the text

et al., 1976) the left radial helix (Figure 5.7). Base 1 presents its hydrophobic surface to a methyl group from Val 119, while base 3 is close to the α-helical main chain between Asp 116 and Ala 117. These two bases stack together (Figure 5.5) and point up into a cavity formed by the left radial helix, a segment of extended chain following the right radial helix, the left radial helix of the 3′ neighbouring subunit, and intersubunit salt-bridges Arg 113–Asp 115 and Arg 112–Asp 88. Base 2 lies along the left radial helix, between the helix and the connecting peptide loop between the left and right slewed helices from the subunit below. Ser 123 and Asn 127 from the top subunit and Asn 33, Gln 34 and Thr 37 from the bottom subunit provide a hydrophilic environment for the polar parts of the base.

Recognition of the origin of assembly sequence need not *a priori* be reflected in the RNA binding; the specificity of the early stages of viral assembly could, for example, be derived from the relative energies of transition states between the unassembled and nucleated states of the growing virus. In fact, however, there do appear to be a number of base-specific hydrogen bonds between the coat protein and the RNA, although specific interactions must not, of course, overwhelm the non-base-specific interactions. The binding site for base 1 is particularly suited to binding guanine (Figure 5.8), thus accounting for the G residues at every third position in the origin of assembly sequence. This base sits between two side-by-side intersubunit ion pairs, Arg 122–Asp 88 and Asp 115–Arg 113, so that if the base is guanine, atom O6 can form a hydrogen bond with Arg 122, and atom N2 with Asp 115. There is less specificity for base 2, although Asn 127 is so placed that it might favour

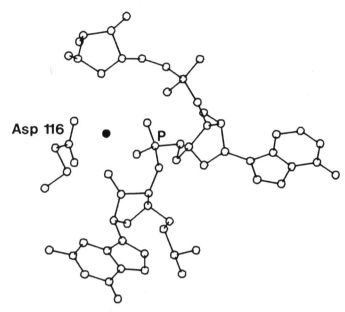

Asp 116

P

Figure 5.6 The phosphate–carboxylate interaction between TMV RNA and Asp 116 of the coat protein. Three nucleotides (with one base omitted for clarity) and the N, C and side chain of Asp 116 are shown. The possible calcium site is marked as a closed circle. The calcium ligands are both oxygens of the phosphate, one carboxylate oxygen, O2′ from ribose 1, O1′ from nucleotide 3, and two water molecules (not shown)

hydrogen bonding with adenine. This Asn side chain could, however, form hydrogen bonds of various degrees of distortion with any base. If base 3 is adenine, N6 can form a hydrogen bond with the main chain carbonyl group of Thr 89. Cytosine could also form this bond, but purines fit the base-binding cavity better. There is thus a strong preference for the sequence GXX, with a possible slight preference for GAA. These observations are consistent with the trinucleoside diphosphate binding results of Steckert and Schuster (1982), although those results are most easily interpreted by assuming that G, when present in the trinucleoside diphosphate, always binds to the base 1 binding site, rather than by assuming that all trinucleoside diphosphates bind in the same phase. With the assumption of strong G binding, the relative binding strengths can be explained in terms of the base specificities described here, and the additional observation that it will always be energetically favourable for the trinucleoside diphosphate to leave the phosphate 2 binding site (close to the negatively charged Asp 116) vacant.

Structural details such as those described here are not yet available for any of the tobamoviruses other than TMV. In sunn-hemp virus and CGMMV-W, GAA does not dominate the triplet pattern at the origin of assembly as it does in TMV (Meshi *et al.*, 1982, 1983). Okada (1986) (see

Protein–Nucleic Acid Interaction

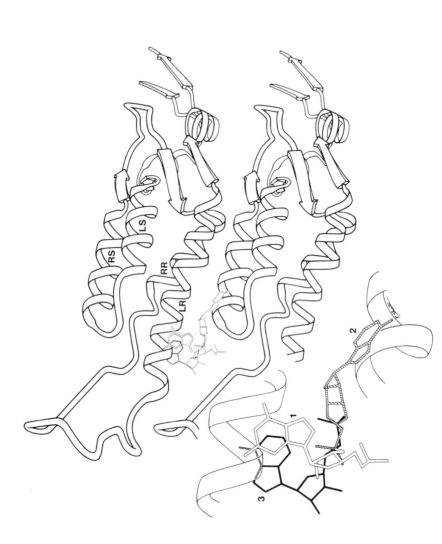

Figure 5.7 Azimuthal view of two subunits of TMV, showing the RNA binding site between them. The viral axis is vertical and to the left of the figure (see Figure 5.1 to orient the subunits in the virus). Four of the α-helices in the protein, the left and right slewed and the left and right radial helices, are marked LS, RS, LR and RR. The inset shows the three nucleotides, drawn as GAA for convenience, more clearly. Nucleotides are labelled 1, 2 and 3 for reference in the text.

also Meshi *et al.*, 1981) favours the target sequence GAXGUUG as the key segment of the origin of assembly loop. Structural studies of CGMMV-W have begun (Lobert *et al.*, 1987), but a complete high-resolution structure determination will be required before it is known whether differences such as these are reflected in the RNA binding sites.

The specificity of TMV coat protein for RNA rather than DNA binding appears to stem from the interactions made by the ribose 2' hydroxyl groups, since all three base-binding sites could accommodate thymine without difficulty. The 2' hydroxyl group from nucleotide 1 is required for two important functions: it forms one of the ligands in the metal binding site, and also forms a hydrogen bond with the non-metal-binding carboxylate oxygen of Asp 116 (Namba *et al.*, 1989). A similar interaction between Asp 115 and a ribose has often been discussed speculatively in the literature, but Asp 115 is more than 8 Å from any ribose hydroxyl. The 2' hydroxyl in ribose 2 stabilizes the close approach of phosphates 2 and 3 by hydrogen bonding to phosphate 3. The ribose 3 hydroxyl makes Van der Waals contacts with Ile 94. These hydroxyl groups might also hydrogen bond to water molecules, not sufficiently well ordered to be visible in a 2.9 Å electron density map, but they do not make hydrogen bonds with the protein. The interactions described here are, however, probably sufficient to account for the protein-RNA specificity.

VIRAL ASSEMBLY AND DISASSEMBLY

Two of the major events of biological interest in TMV, or, indeed, any virus, are the assembly of the virus, in order to prepare it for transport to another host and thus propagate it, and disassembly, the first step in establishing a new infection. These have been studied in TMV for over 50 years; for example, Bawden and Pirie (1937) used glacial acetic acid to extract RNA (although in a somewhat degraded form) from the virus. Assembly has also been of great interest because of TMV's role as a model for self-assembly. It is only very recently, however, that we have obtained a coherent picture of the complete process of assembly and disassembly, and there are still many details yet to be uncovered.

As outlined in Figure 5.3, assembly begins by the interaction between the origin of assembly loop in the RNA and one 20S protein aggregate. Assuming that the 20S aggregate was the disk, Butler *et al.* (1976) suggested that the loop inserts itself from the inside of the disk into the open 'jaws' (Champness *et al.*, 1976) between the two layers of protein subunits. If the nucleating aggregate is the protohelix, however, the RNA binding site would be more difficult to reach (Namba and Stubbs, 1986); the RNA might bind to the bottom surface of the disk, or the disorder in the inner protein loop might still allow it to intercalate between the layers.

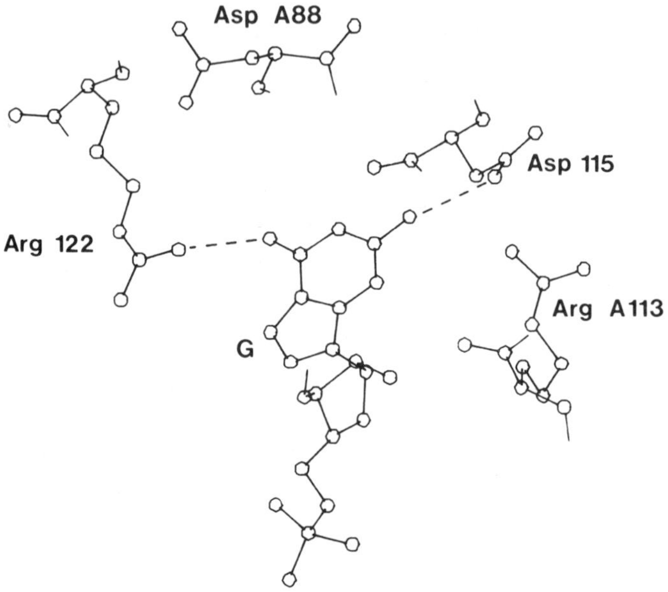

Figure 5.8 Part of the binding site for base 1. Hydrogen bonds between O6 and Arg 122, and between N2 and Asp 115, make the binding of guanine particularly favourable in this site

The initial nucleation is rapidly followed by binding of two more 20S aggregates (Zimmern and Butler, 1977; Zimmern, 1977), perhaps by loose interactions between two additional base-paired stems and loops, equally spaced 75 to 80 nucleotides from the origin of assembly (Zimmern, 1983).

The origin of assembly sequence incorporated into foreign RNAs can permit those RNAs to be encapsidated, forming pseudoviruses (Sleat *et al.*, 1986). This demonstrates that in general, the existence of the sequence is sufficient to allow assembly. Some foreign RNA sequences, however, are not effectively encapsidated, suggesting that there are at least some constraints, perhaps on RNA secondary structure, that limit effective packaging of RNA by TMV coat protein. The importance of having a unique origin of assembly was shown by Gallie *et al.* (1987c), who made long RNA molecules, each containing several copies of the sequence. As would be predicted from the model for the sequence of assembly events, bent and even Y-shaped pseudovirus particles were produced.

After nucleation, elongation of the virions continues by different mechanisms in the 5′ and 3′ directions. There has been considerable controversy about these mechanisms, some of which continues to the present. Some of the points of view have been expressed by Bloomer and Butler (1986), and by Okada (1986). Many of the disagreements are more concerned with states of aggregation of TMV coat protein than with the nature of the protein–RNA interactions, and they will therefore not be discussed in detail here. Some aspects are clear, however.

Elongation in the 5′ direction is considerably faster then in the 3′ direction. Some authors consider these elongation processes to be completely separated in time (for example, Fukuda *et al.*, 1978, although the work described by them was carried out at higher ionic strength than is used by other laboratories), while others observe elongation toward the 3′ end to be slower than, but simultaneous with, elongation in the 5′ direction (for example, Lomonossoff and Butler, 1979). The 5′ process is essentially complete after about 6 minutes. Coat protein can be added to the growing complex, either in the form of the 20S aggregate, or as a 4S aggregate, the A-protein, having about 4 or 5 subunits. The various aggregates of TMV coat protein have received considerable attention, and are discussed in numerous reviews (for example, Butler and Durham, 1977; Stubbs, 1984; Bloomer and Butler, 1986). There is clearly a minimum requirement for the 20S form extending well beyond the need for nucleation; the work of Shire *et al.* (1981) suggests that about 22 per cent of the protein must be from this source. The 20S aggregates could act as an RNA-melting protein, either forming an extended 'nucleation' complex, or simply bridging RNA sequences with high melting energy or low protein affinity wherever they are encountered. Shire *et al.* (1981) found that under the standard conditions used by most laboratories, 48 per cent 20S and 52 per cent 4S protein was incorporated, although Bloomer and Butler (1986) suggest that the fraction of 20S protein incorporated may be rather higher than this.

Little is known about elongation in the 3′ direction. Lomonossoff and Butler (1980) concluded from kinetic experiments that the source of protein is primarily the 4S aggregate, a conclusion which (Figure 5.3) seems topologically most probable. Finally, it is generally recognized that there may be a final stage in assembly, to accommodate end effects associated either with the last turn of the RNA or with the 5′ nucleotide cap.

TMV can be disassembled by a variety of agents, including alkali and detergents. Subunits are progressively lost from the 5′ end (Perham and Wilson, 1976; Wilson *et al.*, 1976), forming a succession of stable intermediates (Perham and Wilson, 1978). There is a particularly stable intermediate one-sixth the length of the virion (Perham, 1969), corresponding to the stage at which the origin of assembly must be uncoated.

It has been known for many years that electrostatic interactions are extremely important in both assembly and disassembly of TMV (reviewed by Caspar, 1963, 1976; Stubbs, 1984). Two carboxyl–carboxylate pairs have been identified (Stubbs *et al.*, 1977; Namba and Stubbs, 1986) as possible sources of electrostatic energy, and, recently, the phosphate–carboxylate interaction described above has also been implicated in the process of disassembly (Namba and Stubbs, 1986; Namba *et al.*, 1989). These concentrations of negative charge have been identified as binding sites for the anomalously titrating protons of TMV (Caspar, 1963; Shalaby and Lauffer, 1977), and as possible calcium binding sites; TMV is known to bind at least two calcium ions per subunit (Gallagher and Lauffer, 1983). A very plausible mechanism for disassembly of TMV, therefore, would be that the low calcium concentration and high pH (relative to extracellular conditions) of the cell could destabilize these electrostatic charges. However, although some slight destabilization of the virus does occur under these conditions, they are not sufficient for disassembly. This problem was resolved by Wilson (1984), who showed that after pretreatment at pH 8.0 (during which process rods did not dissociate detectably), TMV particles could be dissociated *in vitro* by a preparation containing ribosomes. This phenomenon, called cotranslational disassembly, was later observed *in vivo* (Shaw *et al.*, 1986). The process is inhibited in the presence of excess coat protein (Wilson and Watkins, 1986), supporting the theory that the earliest step in infection is the loss of a few protein subunits to expose the RNA to the ribosomes. The inhibition may also explain the phenomenon of cross-protection (McKinney, 1929; Sherwood and Fulton, 1982), in which plants infected by viruses are protected against infection by related viruses.

The early stages of infection, then, might be summarized in the following way. It should be emphasized that although this is a highly plausible sequence of events in view of current knowledge of the virus structure and its disassembly behaviour, it has not been proven by direct observation. When a virion first enters a plant cell, the low intracellular calcium concentration and the relatively high pH remove protons and calcium ions from the two carboxyl–carboxylate pairs and the phosphate–carboxylate calcium site, allowing electrostatic repulsive forces from the remaining negative charges to destabilize the virus. The protein–nucleic acid interactions involving the first 69 residues are weaker than in the rest of the genome because of the absence of guanine bases, so the protein subunits forming about $1\frac{1}{2}$ turns of the virus helix at the 5' end are lost. The first start codon is thus exposed, and ribosomes bind and move toward the 3' end during translation, competing with the coat protein and stripping the rest of the genome. This mechanism not only ensures protection of the genome even under unusual (for example, alkaline) extracellular conditions, but it protects the viral RNA from degradation inside the cells (Shaw *et al.*, 1986).

ACKNOWLEDGEMENTS

I thank Rekha Pattanayek for many useful discussions, and for the use of her research results. This work was supported by grant GM 33265 from the National Institute of Health.

REFERENCES

Abdallah, M. A., Biellmann, J. F., Nordström, B. and Brändén, C.-I. (1975). The conformation of adenosine diphosphoribose and 8-bromoadenosine diphosphoribose when bound to liver alcohol dehydrogenase. *Eur. J. Biochem.*, **50**, 475–481

Ahlquist, P., Strauss, E. G., Rice, C. M., Strauss, J. H., Haseloff, J. and Zimmern, D. (1985). Sindbis virus proteins nsP1 and nsP2 contain homology to nonstructural proteins from several RNA plant viruses. *J. Virology*, **53**, 536–542

Altona, C. and Sundaralingam, M. (1972). Conformational analysis of the sugar ring in nucleosides and nucleotides. A new description using the concept of pseudorotation. *J. Amer. Chem. Soc.*, **94**, 8205–8212

Atabekov, J. G., Novikov, V. K., Vishnichenko, V. K. and Kaftanova, A. S. (1970). Some properties of hybrid viruses reassembled *in vitro*. *Virology*, **41**, 519–532

Bawden, F. C. and Pirie, N. W. (1937). The isolation and some properties of liquid crystalline substances from solanaceous plants infected with three strains of tobacco mosaic virus. *Proc. R. Soc. London, Ser. B*, **123**, 274–320

Beachy, R. N., Zaitlin, M., Bruening, G. and Israel, H. W. (1976). A genetic map for the cowpea strain of TMV. *Virology*, **73**, 498–507

Beachy, R. N. and Zaitlin, M. (1977). Characterization and *in vitro* translation of the RNAs from less-than-full-length, virus-related, nucleoprotein rods present in tobacco mosaic virus preparations. *Virology*, **81**, 160–169

Bernal, J. D. and Fankuchen, I. (1941). X-ray and crystallographic studies of plant virus preparations. *J. Gen. Physiol.*, **25**, 111–165

Bloomer, A. C. and Butler, P. J. G. (1986). Tobacco mosaic virus. Structure and self-assembly. In Van Regenmortel, M. H. V. and Fraenkel-Conrat, H. (eds), *The Plant Viruses*, vol. 2, Plenum, New York, 19–57

Bloomer, A. C., Champness, J. N., Bricogne, G., Staden, R. and Klug, A. (1978). Protein disk of tobacco mosaic virus at 2.8 Å resolution showing the interactions within and between subunits. *Nature*, **276**, 362–368

Bloomer, A. C., Graham, J., Hovmöller, S., Butler, P. J. G. and Klug, A. (1981). Tobacco mosaic virus: interaction of the protein disk with nucleotides and its implications for virus structure and assembly. In Balaban, M., Sussman, J. L., Traub, W. and Yonath, A. (eds), *Structural Aspects of Recognition and Assembly in Biological Macromolecules*, Balaban ISS, Rehovot and Philadelphia, 851–864

Bruening, G., Beachy, R. N., Scalla, R. and Zaitlin, M. (1976). *In vitro* and *in vivo* translation of the ribonucleic acids of a cowpea strain of tobacco mosaic virus. *Virology*, **71**, 498–517

Butler, P. J. G. (1972). Structures and roles of the polymorphic forms of tobacco mosaic virus protein. VI. Assembly of the nucleoprotein rods of tobacco mosaic virus from the protein disks and RNA. *J. Mol. Biol.*, **72**, 25–35

Butler, P. J. G. (1974). Structures and roles of the polymorphic forms of tobacco mosaic virus protein. IX. Initial stages of assembly of nucleoprotein rods from virus RNA and the protein disks. *J. Mol. Biol.*, **82**, 343–353

Butler, P. J. G. (1984). The current picture of the structure and assembly of tobacco mosaic virus. *J. Gen. Virol.*, **65**, 253–279

Butler, P. J. G. and Durham, A. C. H. (1977). Tobacco mosaic virus protein aggregation and the virus assembly. *Adv. Protein Chem.*, **31**, 187–251

Butler, P. J. G. and Klug, A. (1971). Assembly of the particle of tobacco mosaic virus from RNA and disks of protein. *Nature New Biol.*, **229**, 47–50

Butler, P. J. G., Bloomer, A. C., Bricogne, G., Champness, J. N., Graham, J., Guilley, H., Klug, A. and Zimmern, D. (1976). Tobacco mosaic virus assembly – specificity and the transition in protein structure during RNA packaging. In Markham, R. and Horne, R. W. (eds), *Structure–Function Relationships of Proteins*, North Holland, Amsterdam, 101–110

Butler, P. J. G., Finch, J. T. and Zimmern, D. (1977). Configuration of tobacco mosaic virus RNA during virus assembly. *Nature*, **265**, 217–219

Carriquiry, E. and Litvak, S. (1974). Further studies on the enzymatic aminoacylation of TMV-RNA by histidine. *FEBS Lett.*, **38**, 287–291

Caspar, D. L. D. (1963). Assembly and stability of the tobacco mosaic virus particle. *Adv. Protein Chem.*, **18**, 37–121

Caspar, D. L. D. (1976). Switching in the self-control of self-assembly. In Markham, R. and Horne, R. W. (eds), *Structure–Function Relationships of Proteins*, North Holland, Amsterdam, 85–89

Champness, J. N., Bloomer, A. C., Bricogne, G., Butler, P. J. G. and Klug, A. (1976). The structure of the protein disk of tobacco mosaic virus to 5 Å resolution. *Nature*, **259**, 20–24

Correia, J. J., Shire, S., Yphantis, D. A. and Schuster, T. M. (1985). Sedimentation equilibrium measurements of the intermediate-size tobacco mosaic virus protein polymers. *Biochemistry*, **24**, 3292–3297

Cotton, F. A., Hazen, E. E. and Legg, M. J. (1979). Staphylococcal nuclease: proposed mechanism of action based on structure of enzyme-thymidine $3',5'$-bisphosphate-calcium ion complex at 1.5 Å resolution. *Proc. Natl Acad. Sci. USA*, **76**, 2551–2555

Dodds, J. A. and Hamilton, R. I. (1976). Structural interactions between viruses as a consequence of mixed infections. *Adv. Virus Res.*, **20**, 33–86

Einspahr, H. and Bugg, C. E. (1984). Crystal structure studies of calcium complexes and implications for biological systems. In Siegel, H. (ed.), *Metal Ions in Biological Systems*, vol. 17: *Calcium and Its Role in Biology*, Dekker, New York, 51–97

Fraenkel-Conrat, H. and Singer, B. (1964). Reconstitution of tobacco mosaic virus. IV. Inhibition by enzymes and other proteins, and use of polynucleotides. *Virology*, **23**, 354–362

Fraenkel-Conrat, H. and Williams, R. C. (1955). Reconstitution of active tobacco mosaic virus from its inactive protein and nucleic acid components. *Proc. Natl Acad. Sci. USA*, **41**, 690–698

Frist, R. H. (1968). Organized structures assembled *in vitro* from cowpea chlorotic mottle virus protein. *Biophys. J.*, **8**, A69

Fritsch, C., Stussi, C., Witz, J. and Hirth, L. (1973). Specificity of TMV RNA encapsidation: *in vitro* coating of heterologous RNA by TMV protein. *Virology*, **56**, 33–45

Fukuda, M., Ohno, T., Okada, Y., Otsuki, Y. and Takebe, I. (1978). Kinetics of biphasic reconstitution of tobacco mosaic virus *in vitro*. *Proc. Natl Acad. Sci. USA*, **75**, 1727–1730

Fukuda, M., Okada, Y., Otsuki, Y. and Takebe, I. (1980). The site of initiation of rod assembly on the RNA of a tomato and a cowpea strain of tobacco mosaic virus. *Virology*, **101**, 493–502

Fukuda, M., Meshi, T., Okada, Y., Otsuki, Y. and Takebe, I. (1981). Correlation between particle multiplicity and location on virion RNA of the assembly initiation site for viruses of the tobacco mosaic virus group. *Proc. Natl Acad. Sci. USA*, **78**, 4231–4235

Gallagher, W. H. and Lauffer, M. A. (1983). Calcium ion binding by tobacco mosaic virus. *J. Mol. Biol.*, **170**, 905–919

Gallie, D. R., Sleat, D. E., Watts, J. W., Turner, P. C. and Wilson, T. M. A. (1987a). In vivo uncoating and efficient expression of foreign mRNAs packaged in TMV-like particles. *Science*, **236**, 1122–1124

Gallie, D. R., Sleat, D. E., Watts, J. W., Turner, P. C. and Wilson, T. M. A. (1987b). The $5'$-leader sequence of tobacco mosaic virus RNA enhances the expression of foreign gene transcripts *in vitro* and *in vivo*. *Nucleic Acids Res.*, **15**, 3257–3273

Gallie, D. R., Plaskitt, K. A. and Wilson, T. M. A. (1987c). The effect of multiple dispersed copies of the origin-of-assembly sequence from TMV RNA on the morphology of pseudovirus particles assembled *in vitro*. *Virology*, **158**, 473–476

Goelet, P., Lomonossoff, G. P., Butler, P. J. G., Akam, M. E., Gait, M. J. and Karn, J. (1982). Nucleotide sequence of tobacco mosaic virus RNA. *Proc. Natl Acad. Sci. USA*, **79**, 5818–5822

Graham, J. and Butler, P. J. G. (1979). Binding of oligonucleotides to the disk of tobacco-mosaic-virus protein. *Eur. J. Biochem.*, **93**, 333–337

Guilley, H., Jonard, G., Richards, K. E. and Hirth, L. (1975a). Sequence of a specifically encapsidated RNA fragment originating from the tobacco-mosaic-virus coat-protein cistron. *Eur. J. Biochem.*, **54**, 135–144

Guilley, H., Jonard, G., Richards, K. E. and Hirth, L. (1975b). Observations concerning the sequence of two additional specifically encapsidated RNA fragments originating from the tobacco-mosaic-virus coat-protein cistron. *Eur. J. Biochem.*, **54**, 145–153

Guilley, H., Jonard, G., Kukla, B. and Richards, K. E. (1979). Sequence of 1000 nucleotides at the 3′ end of tobacco mosaic virus RNA. *Nucleic Acids Res.*, **6**, 1287–1308

Higgins, T. J. V., Goodwin, P. B. and Whitfield, P. R. (1976). Occurrence of short particles in beans infected with the cowpea strain of TMV. *Virology*, **71**, 486–497

Hirth, L. and Richards, K. E. (1981). Tobacco mosaic virus: model for structure and function of a simple virus. *Adv. Virus Res.*, **26**, 145–199

Holoubeck, V. (1962). Mixed reconstitution between protein from common tobacco mosaic virus and ribonucleic acid from other strains. *Virology*, **18**, 401–404

Hunter, T. R., Hunt, T., Knowland, J. and Zimmern, D. (1976). Messenger RNA for the coat protein of tobacco mosaic virus. *Nature*, **260**, 759–764

IUPAC–IUB Joint Commission on Biochemical Nomenclature (JCBN) (1983). Abbreviations and symbols for the description of conformations of polynucleotide chains. Recommendations 1982. *Eur. J. Biochem.*, **131**, 9–15

Jardetzky, O., Akasaka, K., Vogel, D., Morris, S. and Holmes, K. C. (1978). Unusual segmental flexibility in a region of tobacco mosaic virus coat protein. *Nature*, **273**, 564–566

Keith, J. and Fraenkel-Conrat, H. (1975). Tobacco mosaic virus RNA carries 5′-terminal triphosphorylated guanosine blocked by 5′-linked 7-methylguanosine. *FEBS Lett.*, **57**, 31–33

Lanina, T. P., Terganova, G. V., Ledneva, R. K. and Bogdanov, A. I. (1976). A study of the interaction of TMV protein with single- and double-stranded polynucleotides. *FEBS Lett.*, **67**, 167–170

Lebeurier, G., Nicolaieff, A. and Richards, K. E. (1977). Inside-out model for self-assembly of tobacco mosaic virus. *Proc. Natl Acad. Sci. USA*, **74**, 149–153

Lesyng, B. and Saenger, W. (1981). Influence of crystal packing forces on molecular structure in 4-thiouridine. Comparison of *anti* and *syn* forms. *Z. Naturforsch.*, **C36**, 956–960

Liljas, L. (1986). The structure of spherical viruses. *Prog. Biophys. Molec. Biol.*, **48**, 1–36

Lobert, S., Heil, P. D., Namba, K. and Stubbs, G. (1987). Preliminary X-ray fiber diffraction studies of cucumber green mottle mosaic virus, watermelon strain. *J. Mol. Biol.*, **196**, 935–938

Lomonossoff, G. P. and Butler, P. J. G. (1979). Location and encapsidation of the coat protein cistron of tobacco mosaic virus: a bidirectional elongation of the nucleoprotein rod. *Eur. J. Biochem.*, **93**, 157–164

Lomonossoff, G. P. and Butler, P. J. G. (1980). Assembly of tobacco mosaic virus: elongation towards the 3′-hydroxyl terminus of the RNA. *FEBS Lett.*, **113**, 271–274

McKinney, H. H. (1929). Mosaic diseases in the Canary Islands, West Africa and Gibraltar. *J. Agric. Res.*, **39**, 557–578

Mandelkow, E., Stubbs, G. and Warren, S. (1981). Structures of the helical aggregates of tobacco mosaic virus protein. *J. Mol. Biol.*, **152**, 375–386

Meshi, T., Ohno, T., Iba, H. and Okada, Y. (1981). Nucleotide sequence of a cloned cDNA copy of TMV (cowpea strain) RNA, including the assembly origin, the coat protein cistron, and the 3′ non-coding region. *Mol. Gen. Genet.*, **184**, 20–25

Meshi, T., Ohno, T. and Okada, Y. (1982). Nucleotide sequence and its character of cistron coding for the 30K protein of tobacco mosaic virus (OM strain). *J. Biochem.*, **91**, 1441–1444

Meshi, T., Kiyama, R., Ohno, T. and Okada, Y. (1983). Nucleotide sequence of the coat protein cistron and the 3′ noncoding region of cucumber green mottle mosaic virus (watermelon strain) RNA. *Virology*, **127**, 54–64

Mondragon, A. (1984). X-ray crystallographic refinement of the disk form of tobacco mosaic virus. Ph.D. thesis, University of Cambridge

Moras, D., Olsen, K. W., Sabesan, M. N., Buehner, M., Ford, G. C. and Rossmann, M. G. (1975). Studies of asymmetry in the three-dimensional structure of lobster D-glyceraldehyde-3-phosphate dehydrogenase. *J. Biol. Chem.*, **250**, 9137–9162

Namba, K. and Stubbs, G. (1985). Solving the phase problem in fiber diffraction. Application to tobacco mosaic virus at 3.6 Å resolution. *Acta Cryst.*, **A41**, 252–262

Namba, K. and Stubbs, G. (1986). Structure of tobacco mosaic virus at 3.6 Å resolution: implications for assembly. *Science*, **231**, 1401–1406

Namba, K., Caspar, D. L. D. and Stubbs, G. (1985). Computer graphics representation of levels of organization in tobacco mosaic virus structure. *Science*, **227**, 773–776

Namba, K., Pattanayek, R. and Stubbs, G. (1989). Visualization of protein–nucleic acid interactions in a virus: refined structure of intact tobacco mosaic virus at 2.9 Å resolution by X-ray fiber diffraction. *J. Mol. Biol.*, in press

Öberg, B. and Philipson, L. (1972). Binding of histidine to tobacco mosaic virus RNA. *Biochem. Biophys. Res. Comm.*, **48**, 927–932

Ohno, T., Yamaura, R., Kuriyama, K., Inoue, H. and Okada, Y. (1972). Structure of N-bromo-succinimide-modified tobacco mosaic virus protein and its function in the reconstitution process. *Virology*, **50**, 76–83

Ohno, T., Okada, Y., Shimotohno, K., Miura, K., Shinshi, H., Miwa, M. and Sugimura, T. (1976). Enzymatic removal of the 5'-terminal methylated blocked structure of tobacco mosaic virus RNA and its effects on infectivity and reconstitution with coat protein. *FEBS Lett.*, **67**, 209–213

Ohno, T., Aoyagi, M., Yamanashi, Y., Saito, H., Ikawa, S., Meshi, T. and Okada, Y. (1984). Nucleotide sequence of the tobacco mosaic virus (tomato strain) genome and comparison with the common strain genome. *J. Biochem.*, **96**, 1915–1923

Okada, Y. (1986). Cucumber green mottle mosaic virus. In Van Regenmortel, M. H. V. and Fraenkel-Conrat, H. (eds), *The Plant Viruses*, vol. 2, Plenum, New York, 267–281

Okada, Y., Ohashi, Y., Ohno, T. and Nozu, Y. (1970). Sequential reconstitution of tobacco mosaic virus. *Virology*, **42**, 243–245

Ollis, D. L., Brick, P., Hamlin, R., Xuong, N. G. and Steitz, T. A. (1985). Structure of large fragment of *Escherichia coli* DNA polymerase I complexed with dTMP. *Nature*, **313**, 762–766

Perham, R. N. (1969). Sucrose density-gradient analysis of the alkaline degradation of tobacco mosaic virus. *J. Mol. Biol.*, **45**, 439–441

Perham, R. N. and Wilson, T. M. A. (1976). The polarity of stripping of coat protein subunits from the RNA in tobacco mosaic virus under alkaline conditions. *FEBS Lett.*, **62**, 11–15

Perham, R. N. and Wilson, T. M. A. (1978). The characterization of intermediates formed during the disassembly of tobacco mosaic virus at alkaline pH. *Virology*, **84**, 293–302

Richards, K. E., Guilley, J., Jonard, G. and Hirth, L. (1978). Nucleotide sequence at the 5'extremity of tobacco-mosaic-virus RNA. 1. The noncoding region (nucleotides 1–68). *Eur. J. Biochem.*, **84**, 513–519

Rietveld, K., Linschooten, K., Pleij, C. W. A. and Bosch, L. (1984). The three-dimensional folding of the tRNA-like structure of tobacco mosaic virus RNA. A new building principle applied twice. *EMBO J.*, **3**, 2613–2619

Rochon, D. and Siegel, A. (1984). Chloroplast DNA transcripts are encapsidated by tobacco mosaic virus coat protein. *Proc. Natl Acad. Sci. USA*, **81**, 1719–1723

Saenger, W. (1984). *Principles of Nucleic Acid Structure*, Springer, New York

Saran, A., Pullman, B. and Perahia, D. (1973). Molecular orbital calculations on the conformation of nucleic acids and their constituents. IV. Conformation about the exocyclic $C_{4'}$-$C_{5'}$ bond. *Biochim. Biophys. Acta*, **299**, 497–499

Shalaby, R. A. F. and Lauffer, M. A. (1977). Hydrogen ion uptake upon tobacco mosaic virus protein polymerization. *J. Mol. Biol.*, **116**, 709–725

Shaw, J. G., Plaskitt, K. A. and Wilson, T. M. A. (1986). Evidence that tobacco mosaic virus particles disassemble cotranslationally *in vivo*. *Virology*, **148**, 326–336

Sherwood, J. L. and Fulton, R. W. (1982). The specific involvement of coat protein in tobacco mosaic virus cross protection. *Virology*, **119**, 150–158

Shire, S. J., Steckert, J. J. and Schuster, T. M. (1981). Mechanism of tobacco mosaic virus assembly: incorporation of 4S and 20S protein at pH 7.0 and 20 °C. *Proc. Natl Acad. Sci. USA*, **78**, 256–260

Siegel, A., Hari, V., Montgomery, I. and Kolacz, K. (1976). A messenger RNA for capsid protein isolated from tobacco mosaic virus-infected tissue. *Virology*, **73**, 363–371

Sleat, D. E., Turner, P. C., Finch, J. T., Butler, P. J. G. and Wilson, T. M. A. (1986). Packaging of recombinant RNA molecules into pseudovirus particles directed by the origin-of-assembly sequence from tobacco mosaic virus RNA. *Virology*, **155**, 299–308

Steckert, J. J. and Schuster, T. M. (1982). Sequence specificity of trinucleoside diphosphate binding to polymerized tobacco mosaic virus protein. *Nature*, **299**, 32–36

Stubbs, G. (1984). Macromolecular interactions in tobacco mosaic virus. In Jurnak, F. A. and

McPherson, A. (eds), *Biological Macromolecules and Assemblies*, Vol. I: *Virus Structures*, Wiley, New York, 149–202

Stubbs, G. (1989). Virus structure. In Fasman, G. (ed.), *Prediction of Protein Structure and Principles of Protein Conformation*, Plenum, New York, in press

Stubbs, G. and Stauffacher, C. (1981). Structure of the RNA in tobacco mosaic virus. *J. Mol. Biol.*, **152**, 387–396

Stubbs, G., Warren, S. and Holmes, K. (1977). Structure of RNA and RNA binding site in tobacco mosaic virus from a 4 Å map calculated from X-ray fibre diagrams. *Nature*, **267**, 216–221

Sulzinski, M. A., Gabard, K. A., Palukaitis, P. and Zaitlin, M. (1985). Replication of tobacco mosaic virus. VIII. Characterization of a third subgenomic TMV RNA. *Virology*, **145**, 132–140

Turner, D. R., Mondragon, A., Fairall, L., Bloomer, A. C., Finch, J. T., van Boom, J. H. and Butler, P. J. G. (1986). Oligonucleotide binding to the coat protein disk of tobacco mosaic virus. Possible steps in the mechanism of assembly. *Eur. J. Biochem.*, **157**, 269–274

Wang, A. H.-J., Quigley, G. J., Kolpak, F. J., Crawford, J. L., van Boom, J. M., van der Marel, G. and Rich, A. (1979). Molecular structure of a left-handed double helical DNA fragment at atomic resolution. *Nature*, **282**, 680–686

Wilson, T. M. A. (1984). Cotranslational disassembly of tobacco mosaic virus *in vitro*. *Virology*, **137**, 255–265

Wilson, T. M. A. and Watkins, P. A. C. (1986). Influence of exogenous viral coat protein on the cotranslational disassembly of tobacco mosaic virus (TMV) particles *in vitro*. *Virology*, **149**, 132–135

Wilson, T. M. A., Perham, R. N., Finch, J. T. and Butler, P. J. G. (1976). Polarity of the RNA in the tobacco mosaic virus particle and the direction of protein stripping in sodium dodecyl sulphate. *FEBS Lett.*, **64**, 285–289

Wilson, T. M. A., Perham, R. N. and Butler, P. J. G. (1978). Intermediates in the disassembly of tobacco mosaic virus at alkaline pH. Infectivity, self-assembly, and translational activities. *Virology*, **89**, 475–483

Zimmern, D. (1975). The 5′ end group of tobacco mosaic virus RNA is m⁷GpppGp. *Nucleic Acids Res.*, **2**, 1189–1201

Zimmern, D. (1977). The nucleotide sequence at the origin for assembly on tobacco mosaic virus RNA. *Cell*, **11**, 463–482

Zimmern, D. (1983). An extended secondary structure model for the TMV assembly origin, and its correlation with protection studies and an assembly defective mutant. *EMBO J.*, **2**, 1901–1907

Zimmern, D. and Butler, P. J. G. (1977). The isolation of tobacco mosaic virus RNA fragments containing the origin for viral assembly. *Cell*, **11**, 455–462

Zimmern, D. and Wilson, T. M. A. (1976). Location of the origin for viral reassembly on tobacco mosaic virus RNA and its relation to the stable fragment. *FEBS Lett.*, **71**, 294–298

6

Structural and functional studies of ribonuclease T1

Udo Heinemann and Ulrich Hahn

INTRODUCTION

Ribonuclease (RNase) T1 represents a simple, yet most rewarding model for studying protein–nucleic acid interaction. The isolation of the enzyme from Takadiastase, a commercial preparation of the culture medium of the mould fungus *Aspergillus oryzae*, was first described by Sato and Egami (1957). Since then, RNase T1 has received much attention as a tool in molecular biology, notably for RNA sequencing (Donis-Keller *et al.*, 1977; Simoncsits *et al.*, 1977; Silberklang *et al.*, 1979) and mapping (e.g. Epstein *et al.*, 1981; Nohga *et al.*, 1981; Nomoto *et al.*, 1981; Stackebrandt *et al.*, 1981; Stewart and Crouch, 1981). RNase T1 has also been used to catalyse the formation of phosphodiester bonds in RNA (Podder, 1970).

Apart from its usefulness in the biochemical laboratory, RNase T1 has a number of properties that render the protein a preferred target for investigations by various biophysical techniques. Much of our basic knowledge about RNase T1 has been summarized in the reviews by Egami *et al.* (1980) and by Takahashi and Moore (1982): RNase T1 is a very soluble, heat- and acid-stable monomeric protein of $M_r = 11085$, and displays an isoelectric point close to pH 3. In single-stranded RNA, the enzyme specifically hydrolyses the phosphodiester bonds at the 3'-side of guanosine. The cleavage reaction follows a two-step mechanism where transesterification, to yield a terminal guanosine 2',3'-cyclic phosphate, precedes hydrolysis, which results in a terminal 3'-phosphate. 2'-guanylic acid (2'GMP) is the most potent competitive inhibitor of RNase T1, binding to the protein at pH 5.5 with a dissociation constant of 6.5 μM. Although the base specificity of RNase T1 is different and more pronounced, its mode of RNA cleavage resembles that of bovine pancreatic ribonuclease (Richards and Wyckoff, 1971; Blackburn and Moore, 1982).

The complete covalent structure of RNase T1 (Figure 6.1) has been

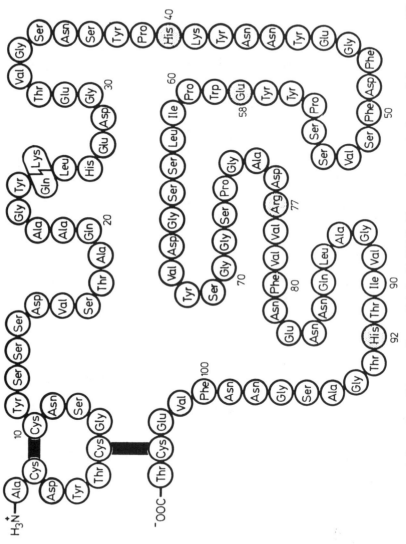

Figure 6.1 Covalent structure of ribonuclease T1, according to Takahashi (1985). The two known isoforms of the enzyme differ at amino acid residue 25. Residues important for the catalytic function of RNase T1 are shown shaded

determined and later revised by Takahashi (1971 and 1985). Some sources of Takadiastase yield a heterogeneous form of RNase T1. The two species present differ in their isoelectric point, but are indistinguishable by their enzymatic activity. The less acidic RNase T1 species has been completely resequenced (Kratzin, unpublished) and shown to contain a lysine residue in place of Gln25 of the protein sequenced by Takahashi. Wherever it is necessary in this text to distinguish between the two isoforms, they will be denoted Lys25-RNase T1 and Gln25-RNase T1, respectively. By a variety of methods, residues His40, Glu58, Arg77 and His92 have been shown to be indispensible for RNase T1 catalytic activity (Egami *et al.*, 1980; Takahashi and Moore, 1982). It will be shown below that these residues are present in all fungal ribonucleases of the RNase T1 family sequenced to date. The single tryptophan residue of RNase T1, Trp59, is located next in sequence to the active site Glu58 and provides a possibility to study motions and binding events by a number of optical techniques.

Since 1982, biochemical investigations aimed at elucidating the origin of base specificity and the catalytic mechanism of RNase T1 have been continued. More importantly, however, several new areas of research have been established. (1) Crystal structure analyses of a number of RNase T1-substrate analogue complexes have revealed the three-dimensional structure of the protein and the architecture of its active site. (2) Recent improvements in instrumentation and experimental, as well as computational, techniques, have made possible time-resolved fluorescence and molecular dynamics studies of RNase T1 that provide information concerning the local environment of the single tryptophan residue and segmental mobility of the protein molecule. In addition, the spatial structure of RNase T1 in solution has been determined by two-dimensional NMR methods. (3) Folding pathways of RNase T1 are being characterized. (4) Genes for RNase T1 have been chemically synthesized and expressed, opening the way to performing site-directed mutagenesis of the protein. (5) A great deal of structural studies on fungal and bacterial ribonucleases related to RNase T1 have been carried out. Comparisons of sequences and tertiary structures of the T1 family ribonucleases have been used to identify structurally and functionally important parts of the proteins. In this chapter, we shall summarize results from these new areas of research. Work covered in the review articles by Egami *et al.* (1980) and Takahashi and Moore (1982) will be considered only when it is directly related to later experimental evidence.

CRYSTALLOGRAPHIC STUDIES OF RIBONUCLEASE T1

During the last decade, RNase T1 and inhibitor complexes thereof have been crystallized in different laboratories under various conditions. Table 6.1 provides a compilation of the crystal forms obtained so far.

Table 6.1   Crystal forms of ribonuclease T1 and RNase T1-substrate analogue complexes

Compound	pH in crystals	Space group	Lattice constants[a]	Crystallographic investigations[b]	Remarks[c]
Gln25-RNase T1	7.2	$P2_12_12_1$	$a = 91.8$ Å $b = 37.4$ Å $c = 77.9$ Å	Cryst.: Martin et al. (1980)	
Lys25-RNase T1* 2'-guanylic acid	4.8–5.2	$P2_12_12_1$	$a = 46.81$ Å $b = 50.11$ Å $c = 40.44$ Å	Cryst.: Heinemann et al. (1980) SD (2.5 Å): Heinemann and Saenger (1982) SR (1.9 Å): Arni et al. (1987, 1988)	MIR $R = 0.180$
Lys25-RNase T1*	6.2–7.2	$P4_32_12$ ($P4_12_12$)	$a = 58.6$ Å $c = 132.8$ Å	Cryst.: Heinemann et al. (1980)	
Lys25-RNase T1* 3',5'-guanosine(bis)phosphate	4.9–5.1	$I23$ ($I2_13$)	$a = 86.47$ Å	Cryst.: Heinemann (1982)	
Gln25-RNase T1* 2'-guanylic acid	4.0	$P2_12_12_1$	$a = 46.65$ Å $b = 50.26$ Å $c = 40.60$ Å	SD (2.4 Å): Sugio et al. (1985a) SR (1.9 Å): Sugio et al. (1988)	MIR $R = 0.203$
Gln25-RNase T1* 3'-guanylic acid	4.0	$P2_12_12_1$	$a = 47.58$ Å $b = 50.92$ Å $c = 40.32$ Å	SR (2.6 Å): Sugio et al. (1985b)	$R = 0.274$
Lys25-RNase T1* Guanylyl 2',5'-guanosine	5.0	$P2_12_12_1$	$a = 47.44$ Å $b = 50.90$ Å $c = 40.43$ Å	SR (1.8 Å): Koepke et al. (1988) in preparation $R = 0.149$	
Lys25-RNase T1* Adenylyl 3',5'-guanosine	5.0	$P2_12_12_1$	$a = 43.4$ Å $b = 50.8$ Å $c = 40.6$ Å	Cryst.: Maslowska (1988)	
Lys25-RNase T1	4.8	$P2_12_12_1$	$a = 46.72$ Å $b = 48.95$ Å $c = 41.34$ Å	Cryst.: Choe et al. (unpublished)	

[a] In cases where slightly varying cell constants for the same crystal form have been quoted in subsequent publications, the numbers were taken from the most recent publication.

[b] Cryst., description of crystallization procedure and determination of crystallographic parameters; SD, structure determination and polypeptide chain tracing; SR, structure refinement. The numbers in parentheses give the resolution of the crystallographic analysis.

[c] MIR, multiple isomorphous replacement. $R = \dfrac{\Sigma |F_O - F_C|}{\ \ }$, the crystallographic residual after refinement.

There are two different crystal forms in space group $P2_12_12_1$ and one each in $P4_12_12$ ($P4_32_12$) and in $I23$ ($I2_13$). The three-dimensional structures of RNase T1-inhibitor complexes determined so far all belong to the same crystal form. To be able to evaluate the possible effects of crystal packing on the molecular conformation, and on the mode of substrate analogue binding by RNase T1, it is highly desirable that the other crystal forms be evaluated in the near future.

The tertiary structure of RNase T1 as present in the complex with 2'GMP (Arni *et al.*, 1987, 1988) is displayed schematically in Figure 6.2. The protein molecule is folded into one compact globular domain, which serves to shield a hydrophobic core, for the most part enclosed between the peripheral α-helix and the central five-stranded antiparallel β-sheet, from contact with the aqueous environment. The amino terminal portion of the polypeptide chain encompasses a separate two-stranded anti-parallel β-sheet and a disulphide bond between residues 2 and 10. A second disulphide bridge between residues 6 and 103 links the carboxy terminus to the amino terminal structure. In between the elements of regular secondary structure, several wide loops are found, which will be shown to be important for the structural and functional integrity of the enzyme.

The complex with 2'-guanylic acid

The three-dimensional structures of crystalline 2'GMP-enzyme complexes have been independently determined in two different laboratories for both Gln25-RNase T1 (Sugio *et al.*, 1985a) and Lys25-RNase T1 (Heinemann and Saenger, 1982). In the latter study, crystals were kept at pH 5, the optimum pH for 2'GMP binding (Takahashi and Moore, 1982), while the Gln25-RNase T1-2'GMP complex was crystallized at pH 4. Crystallographic refinement of both structures (Sugio *et al.*, 1988; Arni *et al.*, 1987, 1988) has revealed only minor differences regarding the conformation of the RNase T1 isoforms and their modes of inhibitor binding. The following discussion will therefore be limited to the crystallographic work carried out in this laboratory.

Our current model of the RNase T1-2'GMP complex is derived from a crystallographic study (Arni *et al.*, 1987), using X-ray diffraction data to a nominal resolution of 1.9 Å (1 Å = 0.1 nm). The structure has been refined by the stereochemically restrained least-squares method of Hendrickson and Konnert (Hendrickson, 1985; Finzel, 1987), supplemented with computer-graphics-aided model building (FRODO; Jones, 1978). At this stage of the analysis, the mean error in atomic coordinates is better than 0.2 Å (Arni *et al.*, 1987). Hence, hydrogen bonds and other interatomic contacts can be assigned with some confidence.

2' GMP binds to a shallow depression on the molecular surface of RNase T1 (Figure 6.3). Most contacts with the guanine base come from

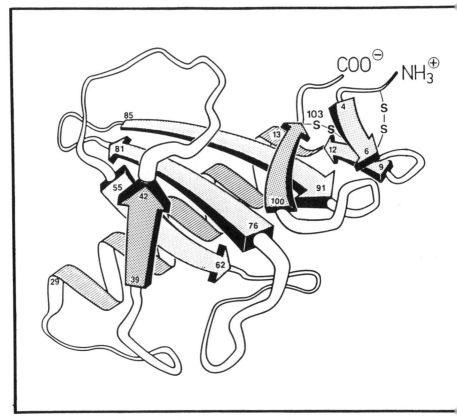

Figure 6.2   Schematic drawing of the tertiary structure of ribonuclease T1 as present in the complex with 2'GMP. The inhibitor molecule has been omitted

amino acid residues 43 to 46, the segment of polypeptide chain following the first strand of the central β-sheet. Since the crucial amino acid residues His40, Glu58, Arg77 and His92 (Egami *et al.*, 1980; Takahashi and Moore, 1982) are all found in the vicinity of the sugar–phosphate portion of 2'GMP, we conclude that the inhibitor has in fact bound to the active site of the enzyme. We may denote the guanosine binding site as the G site and the 2'-phosphate binding site as the p1 site of RNase T1.

In the crystalline complex with RNase T1, 2'GMP adopts a very compact conformation, with *syn* orientation of the glycosyl bond. The exocyclic torsion angle is (+)*gauche* and the sugar pucker C2'-*endo*. This allows a hydrogen bond between the ribose 5'-hydroxyl group and the guanine N3 nitrogen to be formed. The *syn* conformation of the glycosyl bond is not uncommon in purine nucleosides, especially in guanosine, while pyrimidine nucleosides have a marked preference for *anti* (Saenger, 1984). Given the guanine binding to RNase T1 observed in the crystal of the RNase T1-2'GMP complex, the ribose has to be in *syn* to the base in order to

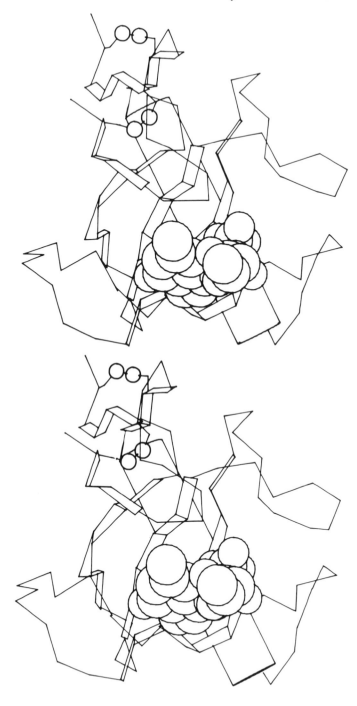

Figure 6.3   Binding of 2'GMP to ribonuclease T1. The inhibitor is given in a space-filling representation. This figure and Figure 6.6 were prepared with a program written by Lesk and Hardman (1982)

expose its 2' and 3' functions to the catalytic site residues.

The specificity of RNase T1 for cleavage after guanosine can be readily explained on the basis of the multiple protein–guanine interactions seen in the crystal structure (Figure 6.4). All hydrogen bonds from the base that can theoretically be formed are actually observed in the G site. The majority of hydrogen bonds connect the guanine ring with the RNase T1 polypeptide backbone, and not with amino acid side chains as might have been expected. The phenolic side chains of Tyr42 and Tyr45 approach the base at oblique angles from both sides to form a sandwich-type arrangement. A regular stacking arrangement, as between bases in nucleic acid helices, is not present between the base and the tyrosine side chains. Since Tyr45 covers the guanine binding site like a lid, there is a possibility that this side chain has to undergo a conformational change upon binding of an RNA chain. It will be interesting to compare the conformation of Tyr45 in inhibitor complexes of RNase T1 and in the free enzyme, the crystal structure of which will, it is hoped, soon be available.

At the catalytic site of RNase T1 we define the immediate environment of the bound inhibitor's sugar–phosphate moiety. While no direct interaction between RNase T1 and the ribose group of 2'GMP is observed, several functional groups of the enzyme form hydrogen bonds to the 2'-phosphate group (Figure 6.5). Among these are the side chains of the essential residues (Oshima *et al.*, 1980; Takahashi and Moore, 1982) His40, Glu58, and, indirectly, Arg77, whose guanidinium group is hydrogen-bonded to the carboxylate of Glu58. Not directly involved is the imidazole group of His92, which is, however, also located in the active site crevice. The phenolic side group of Tyr38 is hydrogen bonded to the 2'-phosphate group as well. Neither this nor any other tyrosine residue of RNase T1 has ever been implicated in the catalytic action of the enzyme.

The complex with guanylyl 2', 5'-guanosine

It has been reported that the kinetics of RNase-T1-catalysed cleavage of GpN substrates differ according to the nature of the leaving group N (Whitfeld and Witzel, 1963; Zabinski and Walz, 1976; Osterman and Walz, 1978; Walz *et al.*, 1979). This would imply the existence of a binding subsite (N1) for the leaving group. In agreement with this hypothesis, further experiments have indicated the binding site of RNase T1 to cover three nucleotides (Osterman and Walz, 1979; Watanabe *et al.*, 1985). The three-dimensional structure of the RNase T1-2'GMP complex offers no clue concerning the course of the polynucleotide chain on the surface of the protein molecule, nor indicates possible binding subsites. Therefore, a crystallographic investigation of guanylyl 2', 5'-guanosine (G(2',5')pG) bound to RNase T1 was carried out (Koepke *et al.*, in preparation). The 2', 5'-phosphodiester linkage of G(2',5')pG is resistant to RNase T1

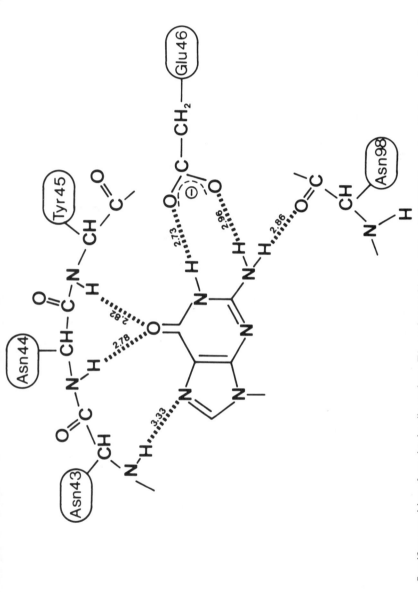

Figure 6.4 Specific recognition of guanine by ribonuclease T1, as observed in the RNase T1*2'GMP complex. Hydrogen bonds are drawn with broken lines and their length is given in Ångstrom units

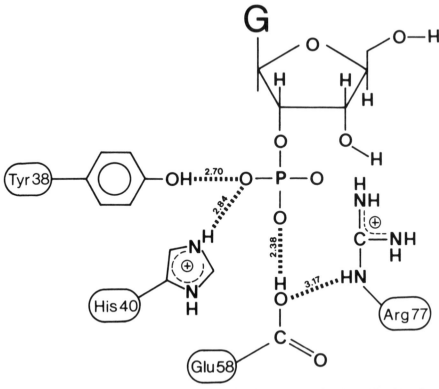

Figure 6.5 Interactions between the inhibitor phosphate group and amino acid residues in the catalytic site of ribonuclease T1. The assignment of hydrogen bonds (broken lines) is tentative, since the crystallographic analysis of the RNase T1*2'GMP complex did not reveal the position of the hydrogen atoms

hydrolysis, and the dinucleotide has been shown to be an effective inhibitor of the enzyme (White *et al.*, 1977).

The complex of RNase T1 with G(2', 5')pG formed crystals isomorphous to RNase T1-2'GMP, and the structure was refined at 1.8 Å resolution to a crystallographic residual of $R = 0.149$ (Koepke *et al.*, in preparation) by methods described for the complex with 2'GMP. The tertiary structure of the protein is nearly identical in both complexes. G(2', 5')pG contains a 5'-proximal guanosine that is bound to the G site of RNase T1 in much the same way as in 2'GMP (Arni *et al.*, 1988).

The 2'-proximal guanosine, which was expected to bind to an N1 subsite of RNase T1, is not bound tightly to the enzyme. The difference electron density was satisfied best when two conformations with relative occupancies of $\frac{2}{3}$ and $\frac{1}{3}$, respectively, were allowed. Figure 6.6 shows the binding to RNase T1 of G(2', 5')pG in the preferred conformation. Only few interactions between protein and the 3'-terminal guanosine are seen, while additional contacts between the inhibitor and symmetry-related protein molecules occur. The structure of the RNase T1-G(2', 5')pG complex therefore does not confirm the proposed existence of an N1 nucleoside binding subsite. It cannot be ruled out, of course, that either the unusual 2',5' linkage or the crystal packing preclude tight binding to this site.

The complex with 3'-guanylic acid

RNase T1 is active only towards internucleoside 3', 5'-phosphodiester linkages and not towards 2', 5'-phosphodiesters. It was thus hoped to gain a better understanding of the active site geometry by co-crystallizing RNase T1 with the reaction product 3'GMP rather than with the inhibitor 2'GMP. The structure of the RNase T1-3'GMP complex was solved and refined at 2.6 Å resolution to $R = 0.274$ (Sugio *et al.*, 1985b). A comparison with these structures is difficult, since neither the resolution of the diffraction data nor the state of the refinement of the RNase T1-3'GMP model, as judged by the crystallographic residual, match the respective values obtained for the complexes described above. Clearly, however, little difference in the overall conformation of RNase T1 exists between all complexes analysed so far.

The guanine base of 3'GMP occupies the same locus in the RNase T1 active site as the 2'GMP base. Unfortunately, the ribose phosphate moiety of 3'GMP is not represented by electron density, owing to disorder. Apparently, the protein–phosphate interactions in the RNase T1-3'GMP complex are insufficient to stabilise a defined conformation of the nucleotide. The reduced affinity to RNase T1 of 3'GMP when compared with 2'GMP (Takahashi and Moore, 1982) can thus be attributed to a tighter binding of the 2'-phosphate group. New data on this issue are expected to come from further refinement of the RNase T1-3'GMP complex, which is in progress (Nishikawa *et al.*, 1987).

A possible mechanism of catalysis

Usher (1969) has described a set of preference rules governing the stereochemistry of the initial transphosphorylation step of ribonucleases. According to these rules, phosphate ester hydrolysis proceeds via a pentacoordinate species with trigonal bipyramidal geometry. In an in-line type reaction, groups enter and leave from the two apical positions of the bipyramid (Figure 6.7); a concerted in-line reaction thus requires simultaneous acid catalysis towards the leaving O5' and base catalysis towards O2' by two functional groups of the enzyme, which must be separated in space by a certain minimum distance. The alternative adjacent reaction cannot be concerted, as it requires pseudorotation of the pentacoordinate intermediate to place the leaving group in an apical position; this reaction type may be catalysed by a single group of the enzyme. In the case of

Figure 6.6 Binding of guanylyl 2′, 5′-guanosine to ribonuclease T1. As described in the text, the inhibitor molecule is present in two alternative conformations in crystals of the RNase T1*G(2′, 5′)pG complex. Here, only the preferred conformation of the inhibitor is shown (space-filling representation)

RNase T1, an in-line reaction is likely to take place, as concluded from the analysis of methanolysis products of 2', 3'-cyclic phosphorothioates (Eckstein *et al.*, 1972).

The arrangement of RNase T1 active site groups in the complex with 2'GMP (Heinemann and Saenger, 1982; Arni *et al.*, 1988) and some simple model building (Heinemann and Saenger, 1983) have led to a proposal for the mechanism of phosphodiester hydrolysis by RNase T1, which is derived from an earlier model published by Takahashi (1970). As shown in Figure 6.8, in transesterification the protonated imidazole group of His92 is proposed to take the role of the general acid, donating a proton to the leaving O5' while Glu58 activates the nucleophilic 2'-hydroxyl group. This mechanism takes account of the active site geometry, the requirement of an in-line reaction type and the known importance for catalysis of the amino acid residues involved (Takahashi and Moore, 1982). It assigns no specific role to Arg77, which may lower the energy of the anionic transition state by virtue of its permanent positive charge, or to His40, which may increase the basicity of Glu58.

The mechanism of RNase T1 catalysed phosphodiester hydrolysis outlined above differs clearly from that proposed by Rüterjans *et al.* (1987), which appears to imply adjacent type stereochemistry. Our reaction scheme bears some resemblance with the two low-pH mechanistic variants of Osterman and Walz (1979), but there are two important differences: clearly, the two histidines have to be interchanged (i.e. His40 and not His92 interacts with Glu58), and the histidine which protonates O5' (His92) has no carboxylate in the immediate vicinity. Recently, the central importance for catalysis of Glu58 has been questioned by Nishikawa *et al.* (1986, 1987, 1988) and Ikehara *et al.* (1987), who propose His40 to be responsible for activation of O2'. Their arguments are derived from the analysis of mutant RNase T1 molecules, and will be discussed under 'site-directed mutagenesis of ribonuclease T1'. Clearly, the crystallographic work performed so far cannot solve the His40/Glu58 riddle.

OTHER BIOPHYSICAL STUDIES OF RIBONUCLEASE T1

Nuclear magnetic resonance spectroscopy

NMR investigations by classical one-dimensional techniques into the structure and function of RNase T1 have continued in the 1980s. With few exceptions, the results are in agreement with the three-dimensional structure and active site geometry of the protein obtained from X-ray crystallography. The interaction with RNase T1 of ^{15}N-enriched 3'GMP was monitored using ^{15}N-NMR (Kyogoku *et al.*, 1982). As expected from the crystallographic studies (Sugio *et al.*, 1985b; Arni *et al.*, 1988; Koepke *et al.*, in preparation), the NMR signals of guanine functions N1, N2 and

IN LINE   ADJACENT

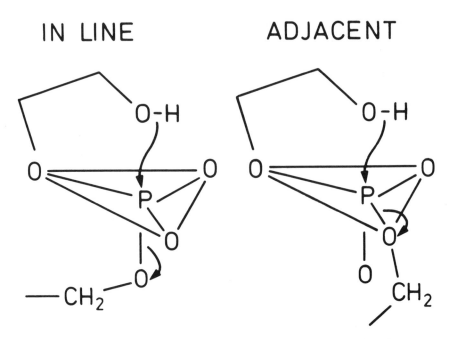

Figure 6.7 The two possible stereochemical routes of phosphodiester hydrolysis, redrawn after Usher (1969). The in-line mechanism requires concerted acid–base catalysis by two functional groups of the enzyme, while the adjacent-type reaction may be catalysed by but one group. In the adjacent reaction, a pseudorotation of the pentacovalent intermediate must occur

N7 were significantly affected upon binding to RNase T1, indicating close proximity to protein groups. ^{1}H-NMR studies of the active site geometry of and inhibitor binding to RNase T1 (Inagaki *et al.*, 1981, 1985) confirm the presence in the active centre of residues His40 and His92, and their interaction with the side chains of Glu58 and Arg77. A finding that is not supported by X-ray crystallography is the proposed vicinity of the His92 imidazole group to a carboxylate function of RNase T1 (Inagaki *et al.*, 1981). In addition, it could be shown that the paramagnetic hexa-cyanochromate anion binds to the active site of RNase T1 as a competitive inhibitor, and can be used to probe the active pocket geometry (Inagaki and Shimada, 1986). By photochemically induced nuclear polarization, followed by peptide analysis, the side chain of Tyr45 could be shown to protrude into the solvent in free RNase T1 and to be in contact with the base in complexes with guanosine nucleotides (Nagai *et al.*, 1985).

A considerable breakthrough has been achieved by the near-complete assignment of the RNase T1 proton resonances and the subsequent determination of the protein's solution structure (Rüterjans *et al.*, 1987). These studies made use of modern two-dimensional NMR procedures (Wüthrich, 1986). The spatial structure of RNase T1 in aqueous solution was determined for the protein alone and for the complexes with 2'GMP and with 3'GMP and refined with molecular dynamics (MD). The most important results of these studies by Rüterjans and colleagues may be summarized as follows: (1) The polypeptide folding in solution is in agreement with the X-ray structure. (2) Substrate analogue binding gives rise to localized conformational changes in the protein. (3) The bound 2'GMP adopts a conformation similar to that in the crystalline complex with RNase T1 (Arni *et al.*, 1988). Since no nuclear Overhauser enhancement (NOE) signals between nucleotide and protein hydrogens could be recorded, little can be said about inhibitor binding.

X-ray crystallography and NMR can be combined to learn more about the structure and function of RNase T1. Where scarcity of non-exchangeable hydrogen atoms (e.g. in nucleotides) or other problems inherent in the NMR technique prevent unambiguous structure determination based on NOEs, information from the crystal structure will be welcome. A practical aspect where NMR information proves useful to the crystallographer, on the other hand, is the orientation of protein side chains. During the refinement of the RNase T1-G(2', 5')pG complex, a number of histidine, glutamine and asparagine side chains were flipped over according to their orientation in the NMR structure (Koepke *et al.*, in preparation). Their conformation would otherwise have remained ambiguous, since at 1.8 Å resolution the X-ray data cannot discriminate carbon from nitrogen or nitrogen from oxygen.

Other spectroscopic studies

The presence of the single tryptophan residue Trp59 has rendered RNase T1 a preferred subject for various spectroscopic techniques. Trp59 is located next in sequence to the active site Glu58 (Takahashi, 1971), but it is not an integral part of the active centre (Heinemann and Saenger, 1982). From spectroscopic studies of the RNase T1, Trp59 information may thus be derived about (1) the microenvironment and (2) the local mobility of the residue, (3) the dynamics of the protein matrix and (4) the binding events at the active site, inasmuch as these influence any of the previous points.

Imakubo and Kai (1977) were the first to show that Trp59 of RNase T1 gives rise to UV-induced phosphorescence in deoxygenated aqueous solution. Their investigations have been extended by Hershberger *et al.* (1980), who employed phosphorescence and optically detected magnetic resonance spectroscopy to RNase T1 to conclude that Trp59 is shielded from the solvent and interacts with a polar group of the protein.

Trp59 has been concluded to be located within the protein matrix from the first fluorescence studies of RNase T1 (Longworth, 1968; Pongs, 1970). Pongs (1970) described an influence on Trp59 fluorescence of 3'GMP

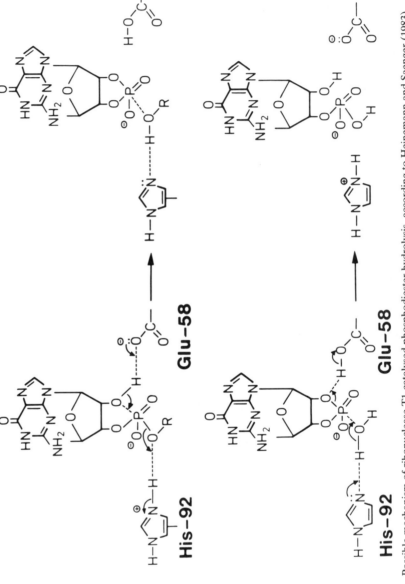

Figure 6.8   Possible mechanism of ribonuclease T1-catalysed phosphodiester hydrolysis, according to Heinemann and Saenger (1983)

binding to RNase T1. Fluorescence titrations of RNase T1 chemically modified at Trp59 showed influences of ionizable groups of the protein, which were assigned to belong to residues Glu58, His40 or His92, and Lys41 (Fukunaga *et al.*, 1982; Fukunaga and Sakiyama, 1982).

Eftink and Ghiron (1975) reported rapid collisional quenching of the Trp59 fluorescence by acrylamide. The conclusion had to be that the protein matrix is sufficiently flexible to permit access of the quenching molecule to the tryptophan side chain. A flexible protein surrounding may allow motions of Trp59 to occur that seem prohibited by the close contacts in the interior of the RNase T1 molecule (Arni *et al.*, 1988). A series of studies using time-resolved fluorescence spectroscopic techniques have attempted to address the problem of motions in RNase T1 (Eftink, 1983; Lakowicz *et al.*, 1983; James *et al.*, 1985; Eftink and Ghiron, 1987; Chen *et al.*, 1987; MacKerell *et al.*, 1987a, 1987b). A clear picture does not emerge from these investigations. There is agreement, however, that at pH values around 5.5 the RNase T1 fluorescence decays monoexponentially or double exponentially, with a major half-life component of 3.6 to 4.0 ns. At neutral pH, the situation becomes more complex (Eftink and Ghiron, 1987; Chen *et al.*, 1987) and very short fluorescence decay times have been reported for substrate analog complexes of RNase T1 (Mackerell *et al.*, 1987a, 1987b).

As motions at the nano- to picosecond time scale become experimentally observable, these data may be compared with MD simulations of RNase T1. MacKerell *et al.* (1987b) have calculated a 55 picosecond trajectory of free RNase T1 *in vacuo*, studies in water are in progress (Rigler, personal communication). The authors find above-average flexibility for a polypeptide segment of RNase T1 including the base binding stretch (amino acids 43–46) and below-average mobility for most residues belonging to the major β-sheet.

RIBONUCLEASE T1 FOLDING

In the very first description of RNase T1 (Sato and Egami, 1957), the heat and acid stability of the protein is noted. These properties have subsequently been exploited in purification protocols for RNase T1 (Uchida and Egami, 1971). The apparent resistance of the enzymatic activity of RNase T1 to heating has later been shown to be due to nearly complete reversibility of the thermal denaturation of the protein (Takahashi, 1974), i.e. the capability of the polypeptide chain to refold to its native conformation after unfolding. Monitoring the Trp59 fluorescence and the peptide group circular dichroism, Oobatake *et al.* (1979a) could show that the thermal denaturation of RNase T1 is a highly co-operative process with a transition temperature of 56 °C at pH 2. Binding of 2'GMP causes an increase in the transition temperature by 6 °C (Oobatake *et al.*, 1979a).

The authors suggested that the denatured state of RNase T1 is not completely random, and also observed some irreversible denaturation of the protein.

RNase T1 is active only when the two disulphide bonds (half-cystines 2–10 and 6–103) are intact (Yamagata *et al.*, 1962). The specific tryptic hydrolysis of the Arg77–Val78 peptide bond in the reduced protein is at least 1700 times faster then in active RNase T1 (Pace and Barrett, 1984), indicating unfolding upon reduction of disulphide bonds. However, Oobatake *et al.* (1979b) could restore partial catalytic activity and native-like folding of disulphide-reduced RNase T1 in the presence of high concentrations of monovalent salts, demonstrating that salt binding stabilizes the active conformation of RNase T1. The interaction of RNase T1 with ions is likely to be specific. Studies of salt-dependent RNase T1 folding can be interpreted in terms of specific binding to the protein of phosphate, plus two monovalent or one divalent cation(s) (Pace and Grimsley, 1988). The location of one calcium binding site on the surface of the protein (Koepke *et al.*, in preparation) corroborates these findings.

The above deals mainly with aspects of the conformational stability of RNase T1 and provides little information as to the pathways of folding. On the other hand, the protein offers possibilities for studying aspects of folding involving the formation of disulphide bridges and *cis/trans* isomerization of the prolyl peptide bonds. Regarding the latter, both Pro39 and Pro55 adopt the *cis* configuration in RNase T1 (Heinemann and Saenger, 1982; Arni *et al.*, 1988). The kinetics of disulphide bond formation in RNase T1 have been described by Pace and Creighton (1986). Proline *cis/trans* isomerization, which has been proposed to determine slow folding steps in several proteins, can be catalysed by a recently discovered prolyl isomerase activity (Lang *et al.*, 1987). First experiments with RNase T1 suggest that folding of this protein is also sensitive to prolyl isomerase (F. X. Schmid, personal communication).

SITE-DIRECTED MUTAGENESIS OF RIBONUCLEASE T1

Chemical gene synthesis and construction of RNase T1 overproducers

Two laboratories have independently started with protein design on RNase T1 (Ikehara *et al.*, 1986; Quaas *et al.*, 1987, 1988a). In both cases, the RNase T1 gene was assembled by the ligation of chemically synthesized oligodeoxyribonucleotides following the approach first introduced by Khorana *et al.* (1972). Ikehara *et al.* (1986) have synthesized the gene for Gln25-RNase T1, whereas Quaas *et al.* (1987, 1988a) have constructed the Lys25-RNase T1 gene. A comparison of the two synthetic genes is shown in Figure 6.9.

The direct expression of RNase T1 genes in *E. coli* failed (Ikehara *et al.*,

```
     1                                  10
Ile   Met   Cys  Tyr  Cys  Ser  Cys  Ser  Ser  Val  Thr
     Phe   Ala  Asp  Thr  Gly  Asn  Tyr  Ser  Asp  Ser

        Sph I
        GCATGCGACTACACTTGCGGTTCTAACTGCTACTCTTCTTCAGACGTTTCTACT
        || <*><*><*>|| <*>||     <*><*><*><*>   || <*><*><*>||
AGATCTTCATGGCTTGCGACTACACCTGCGGCAGCAACTGCTACTCTAGCTCTGACGTTTCTACC
Bgl II


    20            30                          40
    Gln  Ala  TyrLysLeu  Glu  Gly  Thr  Gly  Asn  Tyr  His
Ala  Ala  Gly   Gln   His  Asp  Glu  Val  Ser  Ser  Pro

        Xma III                              Bam HI
GCTCAAGCGGCCGGATATAAACTTCACGAAGACGGTGAAACTGTTGGATCCAATTCTTACCCACAC
<*>|| || || || ||  || <*>|| <*>|| <*>|| <*>|| || || <*><*>|| <*>
GCTCAGGCTGCTGGCTACCAGCTGCACGAGGACGGCGAAACCGTTGGCTCTAACTCTTACCCGCAC
              Pvu II


                      50                    60
    Tyr  Asn  Glu  Phe  Phe  Val  Ser  Tyr  Glu  Pro  Leu
Lys  Asn  Tyr  Gly  Asp  Ser  Ser  Pro  Tyr  Trp  Ile

                              Sst I
AAATACAACAACTACGAAGGTTTTGATTTCTCTGTGAGCTCTCCCTACTACGAATGGCCTATCCTC
<*><*><*><*>|| || || || || || || <*>|| <*><*><*><*>|| <*>||
AAATACAACAACTATGAGGGCTTCGACTTTAGCGTTTCTTCTCCGTACTACGAATGGCCGATCCTG
                                              Gdi II


              70                    80
    Ser  Asp  Tyr  Gly  Ser  Gly  Asp  Val  Phe  Glu  Asn
Ser  Gly  Val  Ser  Gly  Pro  Ala  Arg  Val  Asn  Asn

Xho I                    Xma I
TCGAGCGGTGATGTTTACTCTGGTGGGTCCCCGGGTGCTGACCGTGTCGTCTTCAACGAAAACAAC
|| <*>|| || <*><*>|| <*>||   ||| <*><*><*><*>|| || <*><*><*><*><*>
TCTAGCGGCGACGTTTACTCCGGTGGTAGCCCAGGTGCTGACCGTGTAGTATTCAACGAAAACAAC


              90                  100       104
    Leu  Gly  Ile  His  Gly  Ser  Asn  Phe  Glu  ThrEND
Gln  Ala  Val  Thr  Thr  Ala  Gly  Asn  Val  Cys  ENDEND

                                      Hind III
CAACTAGCTGGTGTTATCACTCACACTGGTGCTTCTGGTAACAACTTCGTTGAATGTACATAAGCTT
|| || <*>|| <*><*>|| <*>||| || <*><*>|| <*><*>|| || <*>|| || <*>  |
CAGCTCGCTGGCGTTATCACCCACACCGGCGCTTCTGGCAACAACTTTGTAGAATGCACCTAATAGTCGAC
              Hae II                                    Sal I
```

Figure 6.9 Synthetic genes coding for ribonuclease T1 of Quaas *et al.* (1988a, b; upper strand) and Ikehara *et al.* (1986; lower strand). Unique restriction sites in both genes are shown above and below the sequences, respectively. The amino acid sequences (above the genes) differ in position 25, as indicated. <*> marks identical codons; | connects identical bases where the codons differ

1986; McKeown, unpublished). This might be due to the cytotoxicity of the ribonuclease on the host cell. To avoid this suicide effect, and also to connect the RNase T1 gene with a well-expressed *E. coli* gene, the construction of a fusion protein was undertaken in both laboratories (Ikehara *et al.*, 1986; Quaas *et al.*, 1988a).

Ikehara's group joined the 139 amino acid codons containing N-terminal part of the chemically synthesized gene for the human growth hormone (hGH-AB) with the RNase T1 gene. As RNase T1 does not contain any methionine, a codon for this amino acid was inserted between the hGH-AB and RNase T1 gene to serve as a cleavage site for liberating the ribonuclease by CNBr treatment from the resulting fusion protein. This fusion protein is not at all or only slightly cytotoxic, although it hydrolyses pGpC to a limited extent (Ikehara *et al.*, 1987).

Quaas *et al.* (1988a) have constructed a fusion protein where the RNase T1 gene is linked with the region coding for the signal peptide of *ompA*, a major outer membrane protein of *E. coli*. In this way, the putatively active ribonuclease is secreted out of the cytoplasm, and the enzyme is liberated from the fusion protein by the host's leader peptidase. Fully active RNase T1, with an N-terminal extension of four amino acids, and, after a deletion mutation, an authentic enzyme, can be isolated routinely from the periplasmic fraction, which contains 15 per cent of RNase T1 with respect to the whole amount of protein. The yield of pure enzyme is about 20 mg per l *E. coli* culture after two purification steps (Quaas *et al.*, 1988b).

Analysis of RNase T1 mutants

In Table 6.2, all published RNase T1 mutants and their catalytic behaviour with respect to GpA or pGpC hydrolysis are listed. In a single-site mutant XnY, the wild type amino acid X (one-letter code) at position n is replaced with amino acid Y. Double mutations are denoted accordingly. It is difficult to compare enzymatic properties, since the mutant proteins have been described with different kinetic parameters.

The RNase T1 mutants, most of which were constructed by Ikehara's group, can be arbitrarily divided into three classes according to substitutions of amino acids being involved in (1) guanine recognition, (2) catalysis or (3) other functions (Ikehara *et al.*, 1986; Nishikawa *et al.*, 1986; Ikehara *et al.*, 1987; Nishikawa *et al.*, 1987; Nishikawa *et al.*, 1988). Mutants concerning amino acids 42–45, which belong to the 'specificity centre', can be compared on the basis of initial velocities. From that point of view, none of the shown substitutions in positions 42, 43 and 45 exerts dramatic effects. The mutant with the lowest value of this series is Y45A, showing 45 per cent activity. In the Y45W mutant, the intial velocity with a value of 142 per cent is markedly higher than that of the wild type. If, on the other hand, asparagine-44 is the site of mutation, v_0 drops below 5 per cent. In structural terms, this observation can be explained by the stabilization of the guanine-binding polypeptide segment through hydrogen bonds between the amide side chain of Asn44 and the backbone as seen in the crystal of the RNase T1-2'GMP complex (Arni *et al.*, 1988). Interestingly, the double mutant N43RN44A has a higher initial velocity than the single mutant N44A, whereas in a comparison of the k_{cat}/K_M values the tendency is reversed.

Table 6.2 RNase T1 mutants

Amino acid position																RNase T1 activity (%)	Ref.	kcat/KM $\text{min}^{-1}\ \mu\text{M}^{-1}$	Ref. a/b	Shorthand
−4	−3	−2	−1	25	40	42	43	44	45	46	58	71	72	73	92					
A	E	F	M	Q	H	Y	N	N	Y	E	E	G	S	P	H	100	a†/c†	2.3/142.5	a/b	Q25 (wt)
				K												100	g†/h†			K25 (wt)
																	f†/h†			108
												P	G	S		0.17	a†			PGS
						F										63	a†/c†	1.6	a	Y42F
									F							77	a†/c†	2.4	a	Y45F
									A							45	c[1]			Y45A
									W							142	c[1]			Y45W
							A									69	a†/c†	1.3	a	N43A
							R									51	a†/c†	0.80	a	N43R
								A								1.3	a†/c†	0.084	a	N44A
							R	A								3	a†/c†	0.065	a	N43RN44A
							H	D								5	a†/c†	0.19	a	N43HN44D§
										A								0.6	e	E46A
					A											0.1/0.01	c‡/d‡	13.7	b	H40A
											D					10	c‡	1.82	b	E58D
											Q					1	c‡			E58Q
											A					7	c‡	7.10	d	E58A
															A	0	c‡/d‡			H92A

<-> amino acids of specificity centre
<*> amino acids of catalytic centre

† Percentage activity based on v_0 (pmole/min/μl).
‡ Not indicated whether the activity was determined as v_0 or k_{cat}/K_M.
§ Analogue to RNase Ms, cleaves also after A, slight specificity shift.
(a) Ikehara et al., 1986; (b) Nishikawa et al., 1986; (c) Ikehara et al., 1987; (d) Nishikawa et al., 1987; (e) Nishikawa et al., 1988; (f) Quaas et al., 1988a; (g) Quaas et al., 1988b; (h) Rao, personal communication.
(wt) wild-type RNase T1 from Aspergillus oryzae.

In the RNase T1 mutant N43HN44D, the base recognition site of RNase Ms, a non-base specific ribonuclease from *Aspergillus saitoi*, has been introduced (see Figure 6.10). Here, the specificity is slightly changed: the mutant shows a higher cleavage rate after adenosine than wild type RNase T1 (Ikehara *et al.*, 1987).

A series of mutations has also been constructed to examine the function of amino acid residues His40, Glu58 and His92 (Nishikawa *et al.*, 1986; Ikehara *et al.*, 1987; Nishikawa *et al.*, 1987), which, besides Arg77, participate in the 'catalytic centre' (Heinemann and Saenger, 1982). As outlined above, hydrolysis of RNA is an acid–base catalysed reaction, where protonated His92 is thought to act as the acid and deprotonated Glu58 as the base. Since Glu58 is supposed to be of central importance for the catalysis, one would expect that replacement of this residue by Ala or Gln should result in inactive ribonuclease. Kinetic data of these mutants with respect to k_{cat}/K_M values contradict this presumption. Compared with the wild type enzyme, the E58A mutant shows 5 per cent and E58Q still 1 per cent activity (Nishikawa *et al.*, 1986), whereas on the other hand a substitution of either His40 or His92 by Ala results in nearly inactive enzyme (Ikehara *et al.*, 1987; Nishikawa *et al.*, 1987).

Ikehara *et al.* (1987) and Nishikawa *et al.* (1987) propose these data to indicate that the mechanism of catalysis in RNase T1 is similar to that in RNase A from bovine pancreas (Wlodawer, 1985), where two histidines are essential for activity. Accordingly, His40 of RNase T1 should take the role of Glu58 in the mechanism depicted in Figure 6.8. The function of Glu58 would then be to provide an 'acidic environment around His40 and to enhance its basicity' (Nishikawa *et al.*, 1987). This mechanism should be considered with caveats, however, because His40 is conserved *only* in the eukaryotic RNases of the T1-family, whereas Glu58 can be found in *all* ribonucleases of this type (see below).

RELATED FUNGAL AND BACTERIAL RIBONUCLEASES

Figure 6.10 shows an alignment of the amino acid sequences of nine fungal and four bacterial ribonucleases, which exhibit a high degree of similarity with RNase T1 and may therefore be named the ribonuclease T1-family. Hartley (1980) has first demonstrated that the sequence of the *Bacillus*

Figure 6.10 Compilation of sequences of microbial ribonucleases forming the RNase T1 family. The sequence alignment is based on that of Hartley (1980). The numbering is for RNase T1, and the horizontal line separates fungal from bacterial enzymes. (G) identifies guanine-specific ribonucleases, enzymes marked (N) are non-specific for RNA cleavage after purines. Invariant residues known to be involved in catalysis are shown in boldface letters, and conserved residues located near the base binding site are represented by italic letters. T1, ribonuclease T1 from *Aspergillus oryzae* (Takahashi, 1985); C2, ribonuclease C2 from *Aspergillus clavatus* (Bezborodova *et al.*, 1983); Ms, ribonuclease Ms from *Aspergillus saitoi* (Watanabe *et al.*, 1982); F1, ribonuclease F1 from *Fusarium moniliforme*

```
                                             '            10            '
T1   (G)                                     A C D Y T C G S N C Y S S - S D V S
C2   (G)                                     D C D Y T C G S H C Y S A - S A V S
Ms   (N)                                 E S C E Y T C G S T C Y W S - S D V S
F1   (G)                                 E S A T T C G S T N Y S A - S Q V R
Pb1  (G)                                 A C A A T C G T V C Y T S - S A I S
Pch1 (G)                                 A C A A T C G S V C Y T S - S A I S
Th1  (G)                                 D T A T C G K V F Y S A - S A V S
U1   (G)                               e G G V S V N C G G T Y Y S S - T Q V N
U2   (R)                         C D I P Q S T N C G G N V Y S N - D D I N
St   (G)                                 Q A P C G D T S G F E Q V R L A D
Sa   (G)                                 D V S G T V C L S A
Ba   (N)       A Q V I N T F D G V A D Y L Q T Y H K L P N D Y I T K S E A Q
Bi   (N)       A V I N T F D G V A D Y L I R Y K R L P N D Y I T K S Q A S

          20            '            30          '             40            '           50
T1   T - - A Q A A G Y Q L H E D G E T V G S N S Y P H K Y N N Y E G F D F
C2   D - - A Q S A G Y Q L E S A G Q S V G R S R Y P H Q Y R N Y E G F N F
Ms   A - - A K A K G Y S L Y E S G D T I - - D D Y P H E Y H D Y E G F D F
F1   A - - A A N A A C Q Y Y Q N D D S A G S T T Y P H T Y N N Y E G F D F
Pb1  S - - A G A A G Y N L Y S T N D D V - - S N Y P H E Y H N Y E G F D F
Pch1 A - - A Q E A G Y D L Y S A N D D V - - S N Y P H E Y R N Y E G F D F
Th1  A - - A S N A A C N Y V R A G S T A G G S T Y P H V Y N N Y E G F R F
U1   R - - A I N N A K - - - - S G Q Y - S S T G Y P H T Y N N Y E G F D F
U2   T - - - - A I Q G A L D D V A N G D R P D N Y P H Q Y Y D - E A S D Q
St   L P P E A T D T Y E L I E K G G P Y P Y P E D G T V F E N R E G I L P
Sa   L P P E A T D T L N L I A S D G P F P Y S Q D G V V F Q N R E S V L P
Ba   A L G W V A S K G N L A D V A - P G - K S I G G D I F S N R E G K L P
Bi   A L G W V A S K G D L A E V A - P G - K S I G G D V F S N R E G R L P

                 '            60          '            70               '
T1   S V S S - - - P Y Y - E W P I L S S G D V Y S G - - - - G S P G A
C2   P V S G - - - N Y Y - E W P I L S S G S T Y N G - - - - G G P G A
Ms   P V G T - - - S Y Y - E Y P I M S D Y D V Y T G - - - - G S P G A
F1   P V D G - - - P Y Q - E F P I K S G G - V Y T G - - - - G S P G A
Pb1  P V F G - - - T Y Y - E F P I L K S C K V Y T G - - - - S S P G A
Pch1 P V S G - - - T Y Y - E F P I L R S G A V Y S G - - - - N S P G A
Th1  K G L S - - - K P F Y E F P I L S S G K T Y T G - - - - G S P G A
U1   S D Y C - D G P Y K - E Y P L K T S S G Y T G - - - - G S P G A
U2   I T L C C G S G P W S E F P L V Y N G P Y Y S S R D N Y V S P G P
St   D C A E - - - G Y Y H E Y T V - - - - K T P S G - - - - D D R G A
Sa   T Q S Y - - - G Y Y H E Y T V - - - - I T P G A - - - - R T R G T
Ba   G K S G - - - R T W R E A D I - - - - N Y T S G - - - F - - R N S
Bi   S A G S - - - R T W R E A D I - - - - N Y V S G - - - F - - R N A

                 80            '            90            '            100
T1   D R V V F N E N N - Q L A G V I T H T G A - S G N N F V E C T
C2   D R V V F N D N D - E L A G L I T H T G A - S G D G F V A C Y
Ms   D R V I F D G D D - E L A G L I T H T G A A G D D F V A C S S S
F1   D R V V I N T N C - E T A G A I T H T G A - S G N N F V G C S N T G
Pb1  D R V I F N D D D - E L A G V I T H T G A - S G N N F V A C T
Pch1 D R V V F N G N D - Q L A G V I T H T G A - S G N N F V A C D
Th1  D R V V I N G Q C - S I A G I I T H T G A - S G D A F V A C G G T
U1   D R V V Y D S N D G T F C G A I T H T G A - S G N N F V Q C S Y
U2   D R V I Y Q T N T G E F C A T V T H T G A A S Y D G F T Q C S
St   R R F V V G - D G G E Y F Y T E D H Y E - - - - - S F R L T I V N
Sa   R R I I C G E A T Q E D Y Y T G D H Y A - - - - - T F S L I D Q T C
Ba   D R I L Y S S D W - L I Y K T T D H Y Q - - - - - T F T K I R
Bi   D R L V Y S S D W - L I Y K T T D H Y A - - - - - T F T R I R
```

(Hirabayashi and Yoshida, 1983); Pb1, ribonuclease Pb1 from *Penicillium brevicompactum* (Shlyapnikov *et al.*, 1984); Pch1, ribonuclease Pch1 from *Penicillium chrysogenum* 152A (Shlyapnikov *et al.*, 1986a); Th1, ribonuclease Th1 from *Trichoderma harzianum* (Polyakov *et al.*, 1987); U1, ribonuclease U1 from *Ustilago sphaerogena* (Takahashi and Hashimoto, 1988); U2, ribonuclease U2 from *Ustilago sphaerogena* (Sato and Uchida, 1975); St, ribonuclease St from *Streptomyces erythreus* (Yoshida *et al.*, 1976); Sa, ribonuclease Sa from *Streptomyces aureofaciens* (Shlyapnikov *et al.*, 1986b); Ba, barnase from *Bacillus amyloliquefaciens* (Hartley and Barker, 1972); Bi, ribonuclease Bi from *Bacillus intermedius* 7p (Aphanasenko *et al.*, 1979) ribonucleases can be matched with fungal RNases, and similarities among these enzymes concerning their tertiary structure have also been pointed out (Hill *et al.*, 1983). In the N-terminal third of the proteins, there is no sequence or spatial structure similarity between the bacterial and fungal sub-groups. RNase U2 is unique in the fungal sub-group, as it contains several insertions and deletions with respect to the proteins from *Aspergillus*, *Fusarium*, *Penicillium* and *Trichoderma*, as well as to RNase U1 from the same source. Among the bacterial proteins, there is a clear divide between the nucleases from *Bacillus* and *Streptomyces*, the latter occupying an intermediate position between pro- and eukaryotic enzymes. Allowing four large insertions in loop regions of the fungal RNases, the *Aspergillus giganteus* ribonuclease α-sarcin (Sacco *et al.*, 1983) and several closely related enzymes can also be aligned with the RNases of the T1 family.

Five amino acid residues are present in all of the enzymes: Glu46, Glu58, Arg77, His92 and Phe100 (RNase T1 numbering). Three of these, Glu58, Arg77 and His92 belong to the catalytic centre of RNase T1, corroborating the catalytic mechanism proposed by Heinemann and Saenger (1983; Figure 6.8). Interestingly, Glu46, which in RNase T1 is involved in specific guanine recognition, is among these conserved residues, although not all RNases included in Figure 6.10 are guanine-specific: RNase U2 cleaves preferentially after purines, while RNase Ms, barnase and binase are mainly unspecific. At present, we are not certain about the role of Glu46 in these ribonucleases, nor about the function of the conserved Phe100.

Regarding the latter, we note that the comparison of ribonuclease crystal structures (Hill *et al.*, 1983) has demonstrated the structural equivalence of the fungal Phe100 with the bacterial Tyr93 (RNase T1 numbering). Since the C-terminal sequences following His92 are rather dissimilar between the fungal and bacterial sub-groups, the alignment around Phe100 as shown in Figure 6.10 may well be in error.

Hill *et al.* (1983) have shown the three-dimensional structures and active site geometries of several microbial ribonucleases to be similar. Since then, a number of crystal structures have been refined and more proteins and enzyme-inhibitor complexes have been crystallized (Table 6.3). It is hoped that this large body of structural data can soon be evaluated to yield more detailed and unambiguous information on the function of the RNase T1 group enzymes.

CONCLUSIONS

We have shown that RNase T1 currently is under attack by investigators using various powerful methods. The interplay of these different experimental techniques and ways of scientific thinking is expected to shed

Table 6.3   X-ray crystallographic investigations of ribonucleases belonging to the RNase T1 family

Source	Ribonuclease[a]	Space group	State of structure analysis[b]	Remarks[c]
Aspergillus clavatus	RNase C2	P2$_1$	SR (1.35 Å): Polyakov et al. (1987a)	R = 0.198
	RNase C2*2'GMP	P2$_1$	SD (2.8 Å): Polyakov et al. (1984)	DF
Penicillium brevicompactum	RNase Pbl	I222	SR (1.4 Å): Pavlovsky et al. (1987)	R = 0.184
	RNase Pbl*3′, 5′GDP	I222	SR (1.24 Å): Pavlovsky et al. (1987)	R = 0.196
Trichoderma harzianum	RNase Th1	P3$_2$21	SR (1.75 Å): Polyakov et al. (1987b)	R = 0.300
Streptomyces erythreus	RNase St	C2	SD (2.5 Å): Nakamura et al. (1982)	MIR
Streptomyces aureofaciens	RNase Sa	P2$_1$2$_1$2$_1$	SR (1.8 Å): Sevcik et al. (1987)	R = 0.172
	RNase Sa*3′GMP	P2$_1$2$_1$2$_1$	SR (1.8 Å): Sevcik et al. (1987)	R = 0.184
Bacillus amyloliquefaciens	Barnase	P3$_2$	SD (2.5 Å): Mauguen et al. (1982)	MIR
Bacillus intermedius 7P	Binase	B2	SD (3.2 Å): Pavlovsky et al. (1983)	MIR

[a] 3′, 5′GDP, 3′, 5′-guanosine(bis)phosphate.
[b] SD, structure determination and polypeptide chain tracing; SR, structure refinement. The numbers in parentheses give the resolution of the crystallographic analysis.
[c] MIR, multiple isomorphous replacement; SIR, single isomorphous replacement; DF, difference Fourier transform. $R = \dfrac{\Sigma |F_O - F_C|}{\Sigma (F_O)}$, crystallographic residual after refinement.

light on the structure and function of this protein and protein–RNA interaction in general. Mutant protein molecules can be generated that can be characterized by X-ray crystallography, NMR, time-resolved fluorescence and other biochemical and biophysical methods. The structures determined in the crystal and in solution will serve as models for folding studies and for MD simulations, and results thereof will create a demand for new mutants and new structure analyses. Considering, in addition, the large body of biochemical data on RNase T1 and related enzymes accumulated over the last thirty years, it may be safely stated that ribonuclease T1 is becoming a prime model case for specific protein–nucleic acid interaction.

ACKNOWLEDGEMENTS

We are grateful to Wolfram Saenger for support and encouragement, to Rainer Quaas, Yannis Georgalis, Hans-Peter Grunert and Murali Rao for helpful discussions, and to Jürgen Koepke, Hui-Woog Choe and Franz X. Schmid for permission to quote unpublished work. Financial support by the Deutsche Forschungsgemeinschaft (Sonderforschungsbereich 9 and Ha 1366/1-1 and 2) is also gratefully acknowledged.

REFERENCES

Aphanasenko, G. A., Dudkin, S. M., Kaminir, L. B., Leshchinskaya, I. B. and Severin, E. S. (1979). Primary structure of ribonuclease from *Bacillus intermedius* 7P. *FEBS Letters*, **97**, 77–80

Arni, R., Heinemann, U., Maslowska, M., Tokuoka, R. and Saenger, W. (1987). Restrained least-squares refinement of the crystal structure of the ribonuclease T1*2'-guanylic acid complex at 1.9 Å resolution. *Acta Cryst.*, **B43**, 548–554

Arni, R., Heinemann, U., Tokuoka, R. and Saenger, W. (1988). Three-dimensional structure of the ribonuclease T1*2'GMP complex at 1.9 Å resolution. *J. Biol. Chem.*, **263**, 15358–15368

Bezborodova, S. I., Khodova, O. M. and Stepanov, V. M. (1983). The complete amino acid sequence of ribonuclease C2 from Aspergillus clavatus. *FEBS Letters*, **159**, 256–258

Blackburn, P. and Moore, S. (1982). Pancreatic ribonuclease. *The Enzymes*, **15**, 317–433

Chen, L. X-Q., Longworth, J. W. and Fleming, G. R. (1987). Picosecond time-resolved fluorescence of ribonuclease T1. *Biophys. J.*, **51**, 865–873

Donis-Keller, H., Maxam, A. M. and Gilbert, W. (1977). Mapping adenines, guanines, and pyrimidines in RNA. *Nucleic Acids Res.*, **4**, 2527–2538

Eckstein, F., Schulz, H. H., Rüterjans, H., Haar, W. and Maurer, W. (1972). Stereochemistry of the transesterification step of ribonuclease T1. *Biochemistry*, **11**, 3507–3512

Eftink, M. R. and Ghiron, C. A. (1975). Dynamics of a protein matrix revealed by fluorescence quenching. *Proc. Natl Acad. Sci. USA*, **72**, 3290–3294

Eftink, M. R. (1983). Quenching resolved fluorescence anisotropy studies with single and multi-tryptophan containing proteins. *Biophys. J.*, **43**, 323–334

Eftink, M. R. and Ghiron, C. A. (1987). Frequency domain measurements of the fluorescence lifetime of ribonuclease T1. *Biophys. J.*, **52**, 467–473

Egami, F., Oshima, T. and Uchida, T. (1980). Specific interaction of base-specific nucleases with nucleosides and nucleotides. *Mol. Biol. Biochem. Biophys.*, **32**, 250–277

Epstein, P., Reddy, R. and Busch, H. (1981). Site-specific cleavage by T1 RNase of U-1 RNA in U-1 ribonucleoprotein particles. *Proc. Natl Acad. Sci. USA*, **78**, 1562–1566

Finzel, B. C. (1987). Incorporation of fast Fourier transforms to speed restrained least-squares refinement of protein structures. *J. Appl. Cryst.*, **20**, 53–55

Fukunaga, Y., Tamaoki, H., Sakiyama, F. and Narita, K. (1982). The role of the single tryptophane residue in the structure and function of ribonuclease T1. *J. Biochem.*, **92**, 143–153

Fukunaga, Y. and Sakiyama, F. (1982). Fluorescence titrations of residue 59 and tyrosine in Kyn 59-RNase T1 and NFK 59-RNase T1. *J. Biochem.*, **92**, 155–161

Hartley, R. W. and Barker, E. A. (1972). Amino-acid sequence of extracellular ribonuclease (barnase) of *Bacillus amyloliquefaciens*. *Nature New Biol.*, **235**, 15–16

Hartley, R. W. (1980). Homology between prokaryotic and eukaryotic ribonucleases. *J. Mol. Evol.*, **15**, 355–358

Heinemann, U., Wernitz, M., Pähler, A., Saenger, W., Menke, G. and Rüterjans, H. (1980). Crystallization of a complex between ribonuclease T1 and 2′-guanylic acid. *Eur. J. Biochem.*, **109**, 109–114

Heinemann, U. (1982). Dreidimensionale Strukturen des Calotropin DI und des Komplexes aus Ribonuclease T1 und Guanosin-2′-monophosphat. Thesis, University of Göttingen

Heinemann, U. and Saenger, W. (1982). Specific protein–nucleic acid recognition in ribonuclease T1-2′-guanylic acid complex: an X-ray study. *Nature*, **299**, 27–31

Heinemann, U. and Saenger, W. (1983). Crystallographic study of mechanism of ribonuclease T1-catalysed specific RNA hydrolysis. *J. Biomol. Struct. Dyn.*, **1**, 523–538

Hendrickson, W. A. (1985). Stereochemically restrained refinement of macromolecular structures. *Methods Enzymol.*, **115**, 252–270

Hershberger, M. V., Maki, A. H. and Galley, W. C. (1980). Phosphorescence and optically detected magnetic resonance studies of a class of anomalous tryptophan residues in globular proteins. *Biochemistry*, **19**, 2204–2209

Hill, C., Dodson, G., Heinemann, U., Saenger, W., Mitsui, Y., Nakamura, K., Borisov, S., Tischenko, G., Polyakov, K. and Pavlovsky, S. (1983). The structural and sequence homology of a family of microbial ribonucleases. *Trends Biochem. Sci.*, **8**, 364–369

Hirabayashi, J. and Yoshida, H. (1983). The primary structure of ribonuclease F1 from *Fusarium moniliforme*. *Biochem. Internat.*, **7**, 255–262

Ikehara, M., Ohtsuka, E., Tokunaga, T. *et al.* (1986). Inquiries into the structure–function relationship of ribonuclease T1 using chemically synthesized coding sequences. *Proc. Natl Acad. Sci. USA*, **83**, 4695–4699

Ikehara, M., Ohtsuka, E., Tokunaga, T. *et al.* (1987). Synthesis and properties of ribonuclease T1 and its mutants. In Bruzik, K. S. and Stec, W. J. (eds), *Biophosphates and Their Analogues – Synthesis, Structure, Metabolism and Activity*, Elsevier, Amsterdam, 335–344

Imakubo, K. and Kai, Y. (1977). Phosphorescence of ribonuclease T1 in solution at 293 K. *J. Phys. Soc. Japan*, **42**, 1431–1432

Inagaki, F., Kawano, Y., Shimada, I., Takahashi, K. and Miyazawa, T. (1981). Nuclear magnetic resonance study on the microenvironments of histidine residues of ribonuclease T1 and carboxymethylated ribonuclease T1. *J. Biochem.*, **89**, 1185–1195

Inagaki, F., Shimada, I. and Miyazawa, T. (1985). Binding modes of inhibitors to ribonuclease T1 as studied by nuclear magnetic resonance. *Biochemistry*, **24**, 1013–1020

Inagaki, F. and Shimada, I. (1986). Hexacyanochromate ion as a paramagnetic anion probe for active sites of enzymes. *J. Inorg. Biochem.*, **28**, 311–317

James, D. R., Demmer, D. R., Steer, R. P. and Verall, R. E. (1985). Fluorescence lifetime quenching and anisotropy studies of ribonuclease T1. *Biochemistry*, **24**, 5517–5526

Jones, T. A. (1978). A graphics model building and refinement system for macromolecules. *J. Appl. Cryst.*, **11**, 268–272

Kanaya, S. and Uchida, T. (1986). Comparison of primary structures of ribonuclease U2 isoforms. *Biochem. J.*, **240**, 163–170

Khorana, H. G., Agarwal, K. L., Büchi, H., Caruthers, M. H., Gupta, N. K., Kleppe, K., Kumar, A., Ohtsuka, E., Raj Bhandary, U. L., van de Sande, J. H., Sgaramella, V., Terao, T., Weber, H. and Yamada, T. (1972). Studies on polynucleotides. CIII. Total synthesis of the transfer ribonucleic acid from yeast. *J. Mol. Biol.*, **72**, 209–217

Kyogoku, Y., Watanabe, M., Kainosho, M. and Oshima, T. (1982). A 15N-NMR study on ribonuclease T1-guanylic acid complex. *J. Biochem.*, **91**, 675–679

Lakowicz, J. R., Maliwal, B. P., Cherek, H. and Balter, A. (1983). Rotational freedom of tryptophan residues in proteins and peptides. *Biochemistry*, **22**, 1741–1752

Lang, K., Schmid, F. X. and Fischer, G. (1987). Catalysis of protein folding by prolyl isomerase. *Nature*, **329**, 268–270

Lesk, A. M. and Hardman, K. D. (1982). Computer-generated schematic diagrams of protein structures. *Science*, **216**, 539–540

Longworth, J. W. (1968). Excited state interactions in macromolecules. *Photochem. Photobiol.*, **7**, 587–596

MacKerell, A. D., Rigler, R., Hahn, U. and Saenger, W. (1987a). Ribonuclease T1: Interaction with 2'GMP and 3'GMP as studied by time-resolved fluorescence spectroscopy. In Ehrenberg, A., Rigler, R., Gräslund, A. and Nilsson, L. (eds), *Structure Dynamics and Function of Biomolecules*, Springer, Berlin, Heidelberg, 260–265

MacKerell, A. D., Jr, Rigler, R., Nilsson, L., Hahn, U. and Saenger, W. (1987b). A time-resolved fluorescence, energetic and molecular dynamics study of ribonuclease T1. *Biophys. Chem.*, **26**, 247–261

Martin, P. D., Tulinsky, A. and Walz, F. G., Jr (1980). Crystallization of ribonuclease T1. *J. Mol. Biol.*, **136**, 95–97

Maslowska, M. (1988). Thesis, Free University of Berlin

Mauguen, Y., Hartley, R. W., Dodson, E. J., Dodson, G. G., Bricogne, G., Chothia, C. and Jack, A. (1982). Molecular structure of a new family of ribonucleases. *Nature*, **297**, 162–164

Nagai, H., Kawata, Y., Hayashi, F., Sakiyama, F. and Kyogoku, Y. (1985). An exposed tyrosine residue of RNase T1 and its involvement in the interaction with guanylic acid. *FEBS Letters*, **189**, 167–170

Nakamura, K. T., Iwahashi, K., Yamamoto, Y., Iitaka, Y., Yoshida, N. and Mitsui, Y. (1982). Crystal structure of a microbial ribonuclease, RNase St. *Nature*, **299**, 564–566

Nishikawa, S., Morioka, H., Fuchimura, K., Tanaka, T., Uesugi, S., Ohtsuka, E. and Ikehara, M. (1986). Modification of Glu58, an amino acid of the active center of ribonuclease T1, to Gln and Asp. *Biochem. Biophys. Res. Commun.*, **138**, 789–794

Nishikawa, S., Morioka, H., Kim, H. J., Fuchimura, K., Tanaka, T., Uesugi, S., Hakoshima, T., Tomita, K., Ohtsuka, E. and Ikehara, M. (1987). Two histidine residues are essential for ribonuclease T1 activity as is the case for ribonuclease A. *Biochemistry*, **26**, 8620–8624

Nishikawa, S., Kimura, T., Morioka, H., Uesugi, S., Hakoshima, T., Tomita, K., Ohtsuka, E. and Ikehara, M. (1988). Glu 46 of ribonuclease T1 is an essential residue for the recognition of guanine base. *Biochem. Biophys. Res. Commun.*, **150**, 68–74

Nohga, K., Reddy, R. and Busch, H. (1981). Comparison of RNase T1 fingerprints of U1, U2, and U3 small nuclear RNA's of HeLa cells, human normal fibroblasts, and Novikoff hepatoma cells. *Cancer Res.*, **41**, 2215–2220

Nomoto, A., Kitamura, N., Lee, J. J., Rothberg, P. G., Imura, N. and Wimmer, E. (1981). Identification of point mutations in the genome of the poliovirus sabin vaccine LSc 2ab, and catalogue of RNase T1- and RNase A-resistant oligonucleotides of poliovirus type 1 (Mahoney) RNA. *Virology*, **112**, 217–227

Oobatake, M., Takahashi, S. and Ooi, T. (1979a). Conformational stability of ribonuclease T1. I. Thermal denaturation and effects of salts. *J. Biochem.*, **86**, 55–63

Oobatake, M., Takahashi, S. and Ooi, T. (1979b). Conformational stability of ribonuclease T1. II. Salt-induced renaturation. *J. Biochem.*, **86**, 65–70

Osterman, H. L. and Walz, F. G., Jr (1978). Subsites and catalytic mechanism of ribonuclease T1: kinetic studies using GpA, GpC, GpG, and GpU as substrates. *Biochemistry*, **17**, 4124–4130

Osterman, H. L. and Walz, F. G., Jr (1979). Subsite interactions and ribonuclease T1 catalysis: kinetic studies with ApGpC and ApGpU. *Biochemistry*, **18**, 1984–1988

Pace, C. N. and Barrett, A. J. (1984). Kinetics of tryptic hydrolysis of the arginine–valine bond in folded and unfolded ribonuclease T1. *Biochem. J.*, **219**, 411–417

Pace, C. N. and Creighton, T. E. (1986). The disulphide folding pathway of ribonuclease T1. *J. Mol. Biol.*, **188**, 477–486

Pace, C. N. and Grimsley, G. R. (1988). Ribonuclease T1 is stabilized by cation and anion binding. *Biochemistry*, **27**, 3242–3246

Pavlovsky, A. G., Borisova, S. B., Strokopytov, B. V., Sanishvili, R. G., Vagin, A. A. and Chepurnova, N. K. (1987). Structure bases for nucleotide recognition by guanyl-specific ribonucleases. In Zelinka, J. and Balan, J. (eds), *Metabolism and Enzymology of Nucleic*

Acids Including Gene Manipulations, vol. 6, Slovak Academy of Sciences, Bratislava, 81–96

Pavlovsky, A. G., Vagin, A. A., Vainstein, N. K., Chepurnova, N. K. and Karpeisky, M. Y. (1983). Three-dimensional structure of ribonuclease from Bacillus intermedius 7P at 3.2 Å resolution. *FEBS Letters*, **162**, 167–170

Podder, S. K. (1970). Synthetic action of ribonuclease T1. *Biochim. Biophys. Acta*, **209**, 455–462

Polyakov, K. M., Vagin, A. A., Tishchenko, G. N. and Bezborodova, S. I. (1984). X-ray structural studies of ribonuclease C2 from *Aspergillus clavatus* and its complex with 2'-GMP. In Zelinka, J. and Balan, J. (eds), *Metabolism and Enzymology of Nucleic Acids Including Gene Manipulations*, vol. 5, Slovak Academy of Sciences, Bratislava, 131–138

Polyakov, K. M., Strokopytov, B. V., Vagin, A. A., Bezborodova, S. I. and Orna, L. (1987a). Three-dimensional structure of RNase C2 from *Aspergillus clavatus* at 1.35 Å resolution. In Zelinka, J. and Balan, J. (eds), *Metabolism and Enzymology of Nucleic Acids Including Gene Manipulations*, vol. 6, Slovak Academy of Sciences, Bratislava, 335–340

Polyakov, K. M., Strokopytov, B. V., Vagin, A. A., Bezborodova, S. I. and Shlyapnikov, S. V. (1987b). Crystallization and preliminary X-ray structural studies of RNase Th1 from *Trichoderma harzianum*. In Zelinka, J. and Balan, J. (eds), *Metabolism and Enzymology of Nucleic Acids Including Gene Manipulations*, vol. 6, Slovak Academy of Sciences, Bratislava, 331–334

Pongs, O. (1970). Influences of pH and substrate analogs on ribonuclease T1 fluorescence. *Biochemistry*, **9**, 2316–2321

Quaas, R., Choe, H-W., Hahn, U., McKeown, Y., Stanssens, P., Zabeau, M., Frank, R. and Blöcker, H. (1987). Protein design – a tool for understanding enzyme action: chemical synthesis of a gene for ribonuclease T1. In Bruzik, K. S. and Stec, W. J. (eds), *Biophosphates and Their Analogues – Synthesis, Structure, Metabolism and Activity*, Elsevier, Amsterdam, 345–348

Quaas, R., McKeown, Y., Stanssens, P., Frank, R., Blöcker, H. and Hahn, U. (1988a). Expression of the chemically synthesized gene for ribonuclease T1 in *Escherichia coli* using a secretion cloning vector. *Eur. J. Biochem.*, **173**, 617–622

Quaas, R., Grunert, H.-P., Kimura, M. and Hahn, U. (1988b). Expression of ribonuclease T1 in *Escherichia coli* and rapid purification of the enzyme. *Nucleosides & Nucleotides*, in press

Richards, F. M. and Wyckoff, H. W. (1971). Bovine pancreatic ribonuclease. *The Enzymes*, **4**, 647–806

Rüterjans, H., Hoffmann, E., Schmidt, J. and Simon, J. (1987). Two-dimensional ^{1}H-NMR investigation of ribonuclease T1 and the complexes of RNase T1 with 2'- and 3'-guanosine monophosphate. In Zelinka, J. and Balan, J. (eds), *Metabolism and Enzymology of Nucleic Acids Including Gene Manipulations*, vol. 6, Slovak Academy of Sciences, Bratislava, 81–96

Sacco, G., Drickamer, K. and Wool, I. G. (1983). The primary structure of the cytotoxin α-sarcin. *J. Biol. Chem.*, **258**, 5811–5818

Saenger, W. (1984). *Principles of Nucleic Acid Structure*, Springer, New York, 76–78

Sato, K. and Egami, F. (1957). Studies on ribonucleases in Takadiastase. *J. Biochem.*, **44**, 753–767

Sato, S. and Uchida, T. (1975). The amino acid sequence of ribonuclease U2 from *Ustilago sphaerogena*. *Biochem. J.*, **145**, 353–360

Sevcik, J., Dodson, E. J., Dodson, G. G. and Zelinka, J. (1987). The X-ray analysis of ribonuclease Sa. In Zelinka, J. and Balan, J. (eds), *Metabolism and Enzymology of Nucleic Acids Including Gene Manipulations*, vol. 6, Slovak Academy of Sciences, Bratislava, 33–45

Shlyapnikov, S. V., Kulikov, V. A. and Yakovlev, G. I. (1984). Amino acid sequence and S–S bonds of *Penicillium brevicompactum* guanyl-specific ribonuclease. *FEBS Letters*, **177**, 246–248

Shlyapnikov, S. V., Bezborodova, S. I., Kulikov, V. A. and Yakovlev, G. I. (1986a). Express analysis of protein amino acid sequences. Primary structure of *Penicillium chrysogenum* 152A guanyl-specific ribonuclease. *FEBS Letters*, **196**, 29–33

Shlyapnikov, S. V., Both, V., Kulikov, V. A., Dementiev, A. A., Sevcik, J. and Zelinka, J. (1986b). Amino acid sequence determination of guanyl-specific ribonuclease Sa from *Streptomyces aureofaciens*. *FEBS Letters*, **209**, 335–339

Silberklang, M., Gillum, A. M. and RajBhandary, U. L. (1979). Use of in vitro 32P labeling in the sequence analysis of nonradioactive tRNAs. *Methods Enzymol.*, **59**, 58–109

Simoncsits, A., Brownlee, G. G., Brown, R. S., Rubin, J. R. and Guilley, H. (1977). New rapid gel sequencing method for RNA. *Nature*, **269**, 833–836

Stackebrandt, E., Ludwig, W., Schleifer, K-H. and Gross, H. J. (1981). Rapid cataloguing of ribonuclease T1 resistant oligonucleotides from ribosomal RNAs for phylogenetic studies. *J. Mol. Evol.*, **17**, 227–236

Stewart, M. L. and Crouch, R. J. (1981). Sensitive and rapid analysis of T1-ribonuclease-resistant oligonucleotides in two-dimensional fingerprinting gels of poliovirus type I genomic RNA. *Analyt. Biochem.*, **111**, 203–211

Sugio, S., Amisaki, T., Ohishi, H., Tomita, K-I., Heinemann, U. and Saenger, W. (1985a). pH-induced change in nucleotide binding geometry in the ribonuclease T1-2'-guanylic acid complex. *FEBS Letters*, **181**, 129–132

Sugio, S., Oka, K-I., Ohishi, H., Tomita, K-I, and Saenger, W. (1985b). Three-dimensional structure of the ribonuclease T1*3'-guanylic acid complex at 2.6 Å resolution. *FEBS Letters*, **183**, 115–118

Sugio, S., Amisaki, T., Ohishi, H. and Tomita, K.-I. (1988). Refined X-ray structure of the low pH form of ribonuclease T1-2'-guanylic acid complex at 1.9 Å resolution. *J. Biochem.*, **103**, 354–366

Takahashi, K. (1970). The structure and function of ribonuclease T1. IX. Photooxidation of ribonuclease T1 in the presence of Rose Bengal. *J. Biochem.*, **67**, 833–839

Takahashi, K. (1971). The structure and function of ribonuclease T1. XV. Amino acid sequence of chymotryptic peptides from performic acid-oxidized and heat-denatured ribonuclease T1 – the complete amino acid sequence of ribonuclease T1. *J. Biochem.*, **70**, 617–634

Takahashi, K. (1974). Effects of temperature, salts, and solvents on the enzymatic activity of ribonuclease T1. *J. Biochem.*, **75**, 201–204

Takahashi, K. and Moore, S. (1982). Ribonuclease T1. *The Enzymes*, **15**, 435–468

Takahashi, K. (1985). A revision and confirmation of the amino acid sequence of ribonuclease T1. *J. Biochem.*, **98**, 815–817

Takahashi, K. and Hashimoto, J. (1988). The amino acid sequence of ribonuclease U1, a guanine-specific ribonuclease from the fungus *Ustilago sphaerogena*. *J. Biochem.*, **103**, 313–320

Uchida, T. and Egami, F. (1971). Microbial ribonucleases with special reference to RNase T1, T2, N1, and U2. *The Enzymes*, **4**, 205–250

Usher, D. A. (1969). On the mechanism of ribonuclease action. *Proc. Natl Acad. Sci. USA*, **62**, 661–667

Walz, F. G., Jr, Osterman, H. L. and Libertin, C. (1979). Base-group specificity and the primary recognition site of ribonuclease T1 for minimal RNA substrates. *Arch. Biochem. Biophys.*, **195**, 95–102

Watanabe, H., Ohgi, K. and Irie, M. (1982). Primary structure of a minor ribonuclease from *Aspergillus saitoi*. *J. Biochem.*, **91**, 1495–1509

Watanabe, H., Ando, E., Ohgi, K. and Irie, M. (1985). The subsite structures of guanine-specific ribonucleases and a guanine-preferential ribonuclease. Cleavage of oligo-inosinic acids and poly I. *J. Biochem.*, **98**, 1239–1245

White, M. D., Rapoport, S. and Lapidot, Y. (1977). Guanylyl 2'-5' guanosine as an inhibitor of ribonuclease T1. *Biochem. Biophys. Res. Commun.*, **77**, 1084–1087

Whitfeld, P. R. and Witzel, H. (1963). On the mechanism of action of Takadiastase ribonuclease T1. *Biochim. Biophys. Acta*, **11**, 338–341

Wlodawer, A. (1985). Structure of bovine pancreatic ribonuclease by X-ray and neutron diffraction. In Jurnak, F. A. and McPherson, A. (eds), *Biological Macromolecules & Assemblies, Vol. 2, Nucleic Acids and Interactive Proteins*, Wiley, New York, 393–439

Wüthrich, K. (1986). *NMR of Proteins and Nucleic Acids*, Wiley, New York

Yamagata, S., Takahashi, K. and Egami, F. (1962). The structure and function of ribonuclease T1. VI. Reduction of disulfide bonds of ribonuclease T1. *J. Biochem.*, **52**, 272–274

Yoshida, N., Sasaki, A., Rashid, M. A. and Otsuka, H. (1976). The amino acid sequence of ribonuclease St. *FEBS Letters*, **64**, 122–125

Zabinsky, M. and Walz, F. G., Jr (1976). Subsites and catalytic mechanism of ribonuclease T1: Kinetic studies using GpC and GpU as substrates. *Arch. Biochem. Biophys.*, **175**, 558–564

7
Tet repressor–*tet* operator interaction

Wolfgang Hillen and Andreas Wissmann

I TETRACYCLINE RESISTANCE DETERMINANTS IN GRAM-NEGATIVE BACTERIA

Tetracycline resistance genes are widespread among gram-negative aerobic bacteria. A large number of independent isolates have been characterized and most of them shown to belong to five different classes of determinants, named classes A to E. They are usually plasmid- or transposon-encoded and share several genetic and mechanistic features (Chopra *et al.*, 1981; Izaki *et al.*, 1966; Levy and McMurray, 1978; Mendez *et al.*, 1980; Marshall *et al.*, 1986; Levy, 1988). Resistance against the drug is achieved by an active export mechanism of tetracycline from the resistant cell. This is mediated by a membrane associated resistance protein (McMurray *et al.*, 1980; Waters *et al.*, 1983; Hillen and Schollmeier, 1983; Nguyen *et al.*, 1983). The expression of this protein is inducible by subinhibitory amounts of tetracycline (Mendez *et al.*, 1980). This regulation of expression occurs at the level of transcription. It is brought about by a repressor protein, which under non-inducing conditions represses expression of the resistance gene, as well as that of its own gene. Induction of tetracycline resistance is achieved by binding of tetracycline to the repressor, thereby inactivating the operator-binding function of the protein and allowing expression of both genes. The regulated genes termed *tetA* for resistance and *tetR* for repressor are arranged adjacent to each other with opposite polarity and share common regulatory sequences comprising at least two promotors and two operators per determinant (Beck *et al.*, 1982; Wray *et al.*, 1981; Jorgensen and Reznikoff, 1979; Hillen *et al.*, 1984; Hillen *et al.*, 1982a; Bertrand *et al.*, 1983; Altenbuchner *et al.*, 1983; Unger *et al.*, 1984a; Unger *et al.*, 1984b; Tovar *et al.*, 1988). A sketch of the genetic properties common to all of these tetracycline resistance determinants is shown in Figure 7.1. Figure 7.2 displays a comparison of the amino acid sequences of the four Tet repressor proteins from the classes A, B, C and D. The protein sequences are quite homologous, which suggests that they share a basically similar tertiary structure. Figure 7.3 shows the recognized *tet* operator DNA sequences originating from the five classes A to E. The operator DNA sequences are also highly conserved, but non-identical. It has been shown that heterologous recognition among these elements is possible with quantitatively different affinities (Klock *et al.*, 1985). Thus, these determinants provide an excellent basis for studying the protein–DNA interactions quantitatively with the goal of identifying the chemical functions in the repressor and operator structures involved in recognition of both components.

II PREPARATION OF TET REPRESSOR PROTEINS AND *tet* OPERATOR CONTAINING DNA FRAGMENTS

The *tetR* genes in the wild type determinants are autogeneously repressed by their own gene products (see references cited above). Thus, the amount of repressor molecules per cell is very limited. Constitutive expression of the Tn*10* (class B) encoded Tet repressor at elevated levels was found to reduce the viability of the respective *E. coli* strains (Klock and Hillen, unpublished). Overproduction of this repressor was achieved when the *tetR* gene was placed under transcriptional control of the phage λ promotor P_L (Oehmichen *et al.*, 1984). The same approach was also successful for the pSC101 (class C) (Unger *et al.*, 1984b), the RA1 (class D) (Tovar and Hillen, manuscript in preparation) and the class E-encoded Tet repressor proteins (Tovar *et al.*, 1988). The Tn*1721*/RP1 (class A)-borne *tetR* gene shows an exceptional feature regarding its expression. Like the cI gene from phage λ, it exhibits identical start nucleotides for transcription and translation (Ptashne *et al.*, 1976; Klock and Hillen, 1986). Unlike the cI situation, however, no evidence for a second upstream promoter has been found. In order to maintain the wild type N-terminus, this protein has been prepared with reduced yield from an *E. coli* strain carrying the autoregulated version of the Tn*1721*-encoded *tetR* gene under inducing conditions in the presence of subinhibitory amounts of tetracycline (Klock and Hillen, 1986). As a result, all five Tet repressor proteins (compare Figure 7.2) are available in at least milligram amounts in electrophoretically homogeneous preparations for *in vitro* studies.

DNA fragments containing the various wild type tandem *tet* operator sequences have been prepared for *in vitro* studies from Tn*10* (Hillen *et al.*, 1982a), Tn*1721*/RP1 (Klock and Hillen, 1986), pBR322 and pSC101 (Unger *et al.*, 1984b), and RA1 (Unger *et al.*, 1984a). Recently, these four *tet* regulatory sequences were subcloned on DNA fragments varying in length between 73 and 81 bp containing *Eco*RI ends in derivatives of the plasmid pVH51 (Hershfield *et al.*, 1976). These constructions have been used to prepare mg amounts of small, pure *tet* DNA fragments containing

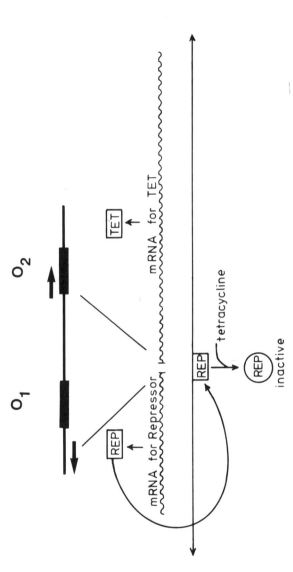

Figure 7.1 Sketch of the genetic and regulatory properties of tetracycline resistance determinants classes A–E. The line denotes the DNA. The wavy lines indicate the two mRNAs with their start points. The rightward mRNA encodes the resistance protein abbreviated TET. The leftward mRNA encodes the Tet repressor called REP. The Tet repressor binds to the *tet* control DNA located between the two mRNA start points and prevents transcription of both genes. The inducer tetracycline binds to the repressor protein and changes its structure such that it does not bind the *tet* control DNA any more. This allows transcription of both genes. The upper part of the figure indicates the location of the *tet* operators O_1 and O_2 with respect to the transcription starts. The two operators are represented by the filled boxes and the arrows indicate the starts of the mRNAs. This basic arrangement is conserved in the class A to E determinants

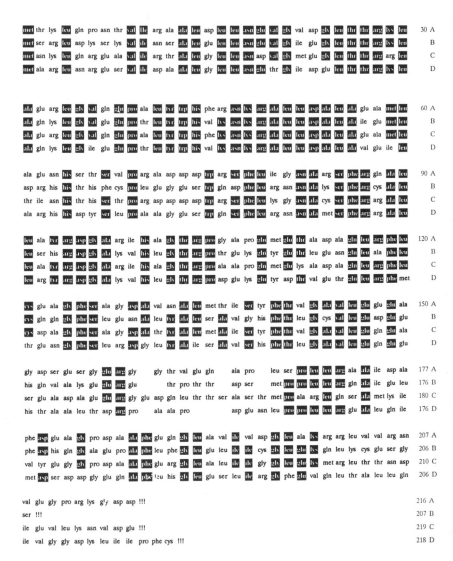

Figure 7.2 Homology among *tetR* sequences from tetracycline resistance determinants. The primary amino acid sequences of the *tetR* genes from tetracycline resistance classes A to D are shown. The spacing of the sequences is aligned to yield maximal homology. Amino acids that are shared by at least three out of the four proteins at a given position are printed white on black. The numbers on the right side denote the positions of the respective amino acids in the primary structures. References to the sequences are given in the text

the four wild type tandem *tet* operator sequences for biophysical studies (Hillen *et al.*, 1981; Tovar and Hillen, to be published).

III TET REPRESSOR–*tet* OPERATOR BINDING

After nucleotide sequence analysis of the Tn*10* encoded *tet* control region (Bertrand *et al.*, 1983), two palindromic sequence elements (Figure 7.3) have been postulated to function as repressor binding sites. This hypothesis has been confirmed by inhibition of restriction at a nearby *Hinc*II site in the presence of Tet repressor (Hillen *et al.*, 1982b), analysis of deletion mutants of the *tet* control sequence *in vivo* (Wray and Reznikoff, 1983), and DNaseI footprinting (Hillen *et al.*, 1984; see also below). DNA fragments containing partial deletions of the Tn*10* *tet* regulatory sequence have been prepared and analysed for Tet repressor binding by the gel mobility shift method (Hillen *et al.*, 1983). This experiment established the stoichiometry of one repressor dimer binding to each of the *tet* operator sequences.

Binding of Tet repressor to *tet* operator stabilizes the *tet*-operator-containing co-operative unit of *tet* DNA against thermal denaturation (Hillen *et al.*, 1982b; Hillen and Unger, 1982), indicating that the protein recognizes both DNA strands of *tet* operator in the double-stranded form. Confirming evidence for the lack of cruciform structures in the palindromic *tet* operator DNA, both alone and in the complex with Tet repressor, was obtained from orientation relaxation studies using electric dichroism at high voltages (Pörschke *et al.*, 1988). The observed relaxation rates were perfectly fitted by models without cruciform structures, which would have resulted in a dramatic decrease in length and much faster rotational relaxation. The same experiment also suggests that no drastic bending of the DNA occurs upon Tet repressor binding. This conclusion has been confirmed by a gel mobility analysis of a set of circular permutated *tet* operator containing DNA fragments complexed with Tet repressor. Although position-dependent differences of gel retardation have been observed in these complexes, the quantitative effect is only about one-tenth of the mobility variation reported for the CRP–DNA complex (Liu-Johnson *et al.*, 1986). The conclusion is, thus, that binding of Tet repressor to *tet* operator results at the most in very small bending of the DNA (Tovar *et al.*, manuscript in preparation). Nevertheless, the binding of Tet repressor to *tet* operator has been shown to induce structural alterations in the DNA. They were measured by circular dichroism changes of the *tet*-operator-containing DNA fragment upon binding of Tet repressor. The nature of this conformational change is not clear. However, the CD change is similar to the one described for a B to A structural transition of DNA (Altschmied and Hillen, 1984).

Each of the five Tet repressor proteins recognizes each *tet* operator

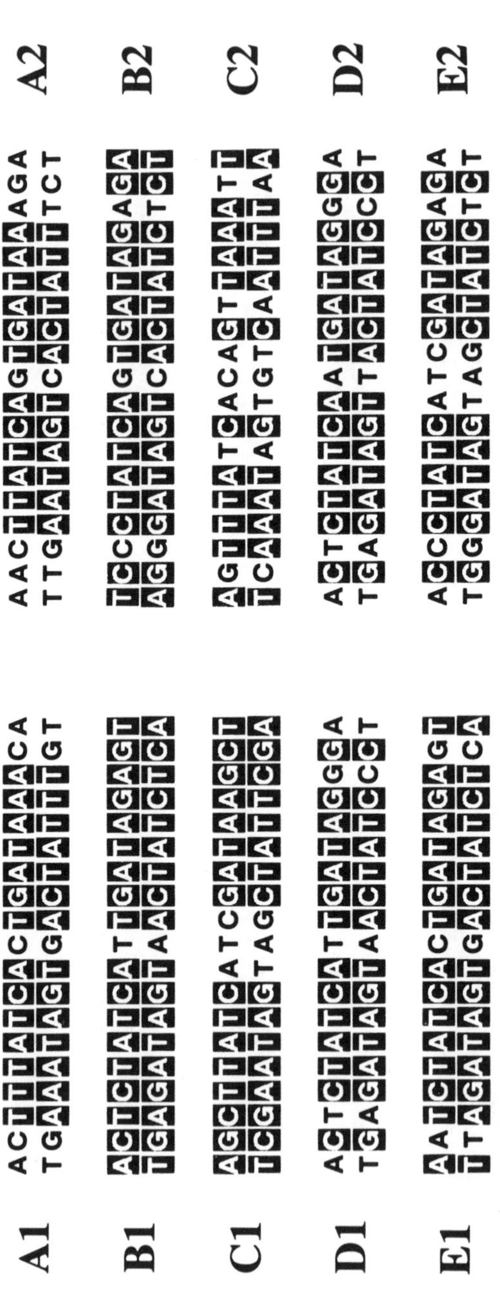

```
A1   ACTTTATCACTGATAAACA        A2   AACTTATCAGTGATAAAGA
     TGAAATAGTGACTATTTGT             TTGAATAGTCACTATTTCT

B1   ACTCTATCATTGATAGAGT        B2   TCCCTATCAGTGATAGAGA
     TGAGATAGTAACTATCTCA             AGGGATAGTCACTATCTCT

C1   AGCTTATCATCGATAAGCT        C2   AGTTTATCACAGTTAAATT
     TCGAATAGTAGCTATTCGA             TCAAATAGTGTCAATTTAA

D1   ACTCTATCATTGATAGGGA        D2   ACTCTATCAATGATAGGGA
     TGAGATAGTAACTATCCCT             TGAGATAGTTACTATCCCT

E1   AATCTATCACTGATAGAGT        E2   ACCCTATCATCGATAGAGA
     TTAGATAGTGACTATCTCA             TGGGATAGTAGCTATCTCT
```

Figure 7.3   Comparison of the *tet* operator sequences from five tetracycline resistance determinants. The *tet* operator sequences from the five classes A to E of tetracycline resistance determinants are shown. The *tet* operators O_1 and O_2 of each class are designated A1, A2, etc., respectively. The palindromic base pairs within each operator are printed white on black

sequence-specifically compared with random non-specific DNA. A detailed study confirms this result for all sixteen possible interactions among the regulatory elements of classes A to D (Tovar *et al.*, manuscript in preparation). However, the affinity constants for these sixteen recognition reactions vary over at least five orders of magnitude (see also below). A comparison of the interaction of the Tn*10*-encoded *tet* operator sequence with Tet repressor proteins from classes A–D studied by the gel mobility shift method is presented in Figure 7.4. At the concentrations employed in this experiment, all four Tet repressors recognize the Tn*10 tet* operator, showing that the respective association constants are greater than 10^7 M^{-1}. Under these salt conditions, non-specific binding occurs with an association constant smaller than 10^6 M^{-1} (compare also below).

The tandem arrangement of the *tet* operators immediately triggers the idea of co-operative binding of two Tet repressor dimers to these sites. Initial experiments had indeed indicated co-operative binding (Hillen and Unger, 1982). Co-operativity is well established for the binding of λ repressor to the λ operators (Hochschild and Ptashne, 1986). The λ operators O_{R1} and O_{R2} are 24 bp apart (Ptashne *et al.*, 1980), whereas the *tet* operators are between 28 and 32 bp apart (Klock *et al.*, 1985). This indicates that, with respect to the λ operator situation, the *tet* operators are roughly one half-turn of the DNA helix further apart. Thus, if Tet repressor-*tet* operator binding would be co-operative it could not simply be modelled in analogy to the λ situation because the proteins would be located next to each other on nearly the same side of the DNA helix. All quantitative interpretations of Tn*10*-Tet repressor–Tn*10-tet* operator interaction (compare below) were obtained without considering co-operativity of binding to the tandem *tet* operator arrangement (Kleinschmidt *et al.*, 1988). *In vivo* studies of differential regulation exerted by the *tet* operators O_1 and O_2 on *tetA* and *tetR* expression also support this notion (Meier *et al.*, 1988). Despite the earlier speculation (Hillen and Unger, 1982), co-operativity may not be important for homologous Tet repressor–*tet* operator interaction.

The titration experiments displayed in Figure 7.4 reveal different results with respect to binding for heterologous Tet repressor–*tet* operator recognition. This is judged by the intensity of the band corresponding to the half-saturated DNA, which is clearly seen in the Tn*10* Tet repressor–Tn*10 tet* operator titration. It is less intense in the RA1 Tet repressor–Tn*10 tet* operator experiment and it is completely lacking in the remaining two titrations. It appears as if the homologous interaction is non-co-operative. Possible explanations other than a co-operative binding mode for the reduction of half-saturated complexes in the heterologous reactions are presently under study.

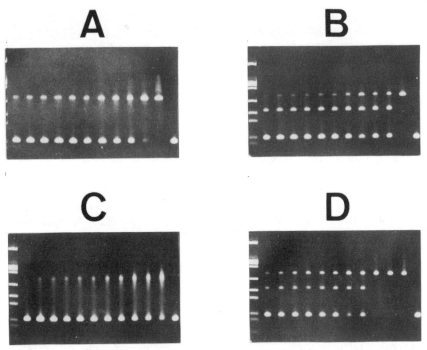

Figure 7.4   Titration of the Tn*10*-encoded *tet* control DNA with class A to D Tet repressor proteins. Titrations of the Tn*10*-encoded *tet* control fragments with the Tet repressor proteins from classes A (Tn*1721*, designated A), B (Tn*10*, designated B), C (pSC101, designated C) and D (RA1, designated D) are shown. The reaction mixtures were analysed by electrophoresis in 5 per cent polyacrylamide gels, as described (Hillen *et al.*, 1984). The total concentration of DNA in each lane is 9.5×10^{-7} M. The rightmost lanes in each gel contain the free 76 bp DNA fragment. The following lanes to the left contain decreasing amounts of the repressor proteins. The repressor concentrations given in 10^{-7} M dimer are, from right to left, in gel A (0; 24; 12; 8; 6; 4.8; 4.0; 3.4; 3.0; 2.6; 2.4; 1.2), in gel B (0; 22; 11; 7.3; 5.5; 4.4; 3.7; 3.1; 2.7; 2.4; 2.2; 1.1) in gel C (0; 21; 10.5; 7.0; 5.2; 4.2; 3.5; 3.0; 2.6; 2.3; 2.1; 1.1), and in gel D (0; 30; 15; 10; 7.5; 6.0; 5.0; 4.3; 3.8; 3.3; 3.0; 1.5). Saturation with the different Tet repressor proteins is defined by the complete titration of the DNA obtained in (A) the third lane from the right; (B) the second lane; ((C): no saturation obtained) and (D) the fourth lane

IV   THERMODYNAMICS AND KINETICS OF *tet* OPERATOR–TET REPRESSOR INTERACTIONS

The Tn*10*-encoded Tet repressor contains two tryptophan residues in its primary structure (Postle *et al.*, 1984), giving rise to typical protein fluorescence. Since this spectroscopic property of the repressor is sensitive to ligand binding, it has been used to study its interaction with the inducer tetracycline and with *tet* operator quantitatively in solution (Takahashi *et al.*, 1986; Kleinschmidt *et al.*, 1988). Non-specific binding of DNA is mostly entropy driven, resulting from the release of 3 to 4 sodium ions. In contrast, 7 to 8 sodium ions are released upon formation of the sequence-specific Tet repressor–*tet* operator complex. Non-electrostatic interactions contribute an additional free energy term of -33 kJ/mol (Kleinschmidt *et al.*, 1988). The dynamics of Tet repressor–*tet* operator association reveal a two-step process with a fast bimolecular association constant of 3×10^8 $M^{-1} s^{-1}$ and a monomolecular second reaction with a rate constant of $6 \times 10^4 s^{-1}$. The first depends on the ionic strength and is attributed to non-specific binding, while the second step is almost independent of the salt concentration and is thought to reflect the formation of the specific complex. Furthermore, the association of Tet repressor–*tet* operator complex formation is almost independent of the length of the DNA fragment which indicates only very little – if any – contribution of sliding. This result is clearly different from the one obtained for Lac repressor–*lac* operator interaction, where extensive sliding seems to contribute to operator recognition (see Berg and von Hippel, 1985).

It is somewhat difficult to compare the binding constants of Tet repressor for *tet* operator and for the inducer tetracycline, because they were measured under different salt conditions (Takahashi *et al.*, 1986; Kleinschmidt *et al.*, 1988). However, in contrast to other repressor–inducer interactions, the Tet repressor–tetracycline complex exhibits a high equilibrium association constant of $10^9 M^{-1}$. This renders induction of transcription of the *tet* genes very sensitive to small amounts of the drug, resulting in a very efficient genetic switch. This property is clearly necessary for resistance, because the inducer is an inhibitor of translation. It binds to ribosomes with an association constant of about $10^6 M^{-1}$ (Epe and Woolley, 1984). Thus there is a concentration range of three orders of magnitude for efficient induction of resistance.

V DNA BINDING MOTIF OF TET REPRESSOR

The primary structures of many procaryotic DNA binding proteins, including the Tet repressors, show homology in regions assumed to form an α-helix–turn–α-helix structural motif which mediates sequence specific recognition of DNA (see Pabo and Sauer, 1984, for a review). The analysis of *trans* dominant *tetR* mutations which show reduced affinity to *tet* operator revealed that five out of seven single amino acid exchanges are clustered within the proposed α-helix–turn–α-helix motif (Isackson and Bertrand, 1985). These mutants affect only the ability of Tet repressor to bind *tet* operator, whereas no reduction in affinity for tetracycline is observed. The remaining two mutations involve the replacement of Gly by Glu and affect the binding affinity for both ligands of Tet repressor. The *trans* dominance of these mutations excludes a totally changed tertiary structure as a reason for their reduced affinity for *tet* operator, because it had been shown that the active form of the native Tet repressor is a dimer

(Hillen *et al.*, 1983). Figure 7.5 shows the primary structures of the α-helix–turn–α-helix motif of four Tet repressor sequences in comparison to the respective parts of the λ cro and λ cI primary structures (Ohlendorf *et al.*, 1983; Sauer *et al.*, 1982).

The α-helix of Tet repressor assumed to make sequence specific contacts with DNA contains a tryptophan residue (Postle *et al.*, 1984). Using oligonucleotide directed mutagenesis to replace a second Trp in the primary structure of Tet repressor by Phe, this was made a single fluorescence probe. The resulting mutant Tet repressor had properties nearly identical to those of the wild type protein (Hansen and Hillen, 1987; Hansen *et al.*, 1987). Accessibility studies with iodide indicated that the Trp in the recognition α-helix is maximally accessible for this quencher of fluorescence. It was therefore concluded that this α-helix is solvent exposed on the surface of the free repressor protein (Hansen *et al.*, 1987). When the inducer tetracycline is bound, the accessibility of this Trp is markedly reduced. In this state, the repressor is not able to bind the *tet* operator sequence-specifically. This result indicates that the recognition α-helix moves somewhat towards the interior of the protein upon binding of the inducer rendering this domain less accessible for DNA. It has been shown by X-ray analysis that the position of the respective α-helix in Trp repressor is slightly different between the operator binding Trp repressor–tryptophan complex and the non-binding aporepressor (Zhang *et al.*, 1987). It is thus quite feasible that the difference in accessibility for iodide detected by fluorescence results from the molecular switch mediating gene regulation in the Tet repressor protein.

Several repressor proteins, including the lac (Weber and Geisler, 1978), λ cI (Roberts and Roberts, 1975), and lex (Little *et al.*, 1980) repressors show domain structures in which the DNA binding functions are located in one domain, while the oligomerization functions are located in a different domain, which can be prepared by limited proteolysis or autolysis of the respective protein. Attempts to produce stable fragments of Tet repressor by limited proteolysis have failed (Heuer and Hillen, unpublished). Furthermore, thermal denaturation profiles of Tet repressor indicate that the entire protein denatures co-operatively in a single process (Wagen-höfer *et al.*, 1988). These observations suggest that Tet repressor might not have a pronounced domain structure. In this respect, it seems to be similar to the Trp repressor (Schevitz *et al.*, 1985).

VI   CONTACT SITES OF THE *tet* OPERATOR DNA WITH TET REPRESSOR

The interaction of Tet repressor with *tet* operator has been studied by a variety of methods, the results of which are summarized in Figure 7.6. DNaseI footprinting has revealed the extension of Tet repressor on the *tet*

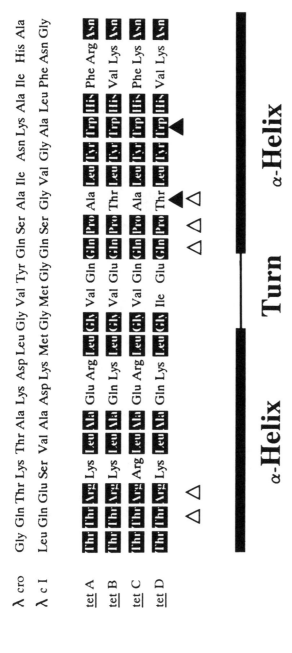

Figure 7.5 Primary structures of the operator binding motifs of four Tet repressors. The amino acid sequences of four proposed DNA binding motifs of Tet repressors from classes A to D are shown in comparison with the respective sequences of the λ *cro* and *c*I repressors. Identical amino acids within the four Tet repressors are printed white on black. The α-helix–turn–α-helix assignment on the bottom of the figure is based upon analogy to the respective motifs of the λ *cro* and *c*I repressors. The open triangles indicate positions important for operator binding defined by mutations in the Tn*10*-encoded Tet repressor (Isackson and Bertrand, 1985). Filled triangles define contacts to the *tet* operator DNA (see text for details)

operator DNA (Hillen *et al.*, 1984). A Tet repressor dimer covers between 22 and 25 bp at the most, as taken from the DNaseI protection. The palindromic sequence is 19 bp long. It is quite possible that only these are covered because DNaseI is assumed to require some space on the DNA for the cleavage reaction. Non-base-specific contacts of Tet repressor to *tet* operator DNA were detected by ethylation interference studies of repressor binding after modification of DNA with *N*-ethylnitroso urea (Heuer and Hillen, 1988). They establish eight essential and four somewhat weaker contacts of a Tet repressor dimer to the *tet* operator. The positions of these contacts are also indicated in Figure 7.6. These sites are located on the same side of the B-DNA double helical structure of *tet* operator.

Phosphate contacts are thought to contribute to the positioning of the recognition α-helix on the operator. In comparison with other well-studied protein–DNA recognition events (see Ebright, 1986, for a detailed discussion), the phosphate contacts would predict the sequence TCTATC in three cases and CCTATC in one case (O_2) as the contact sites for the recognition α-helix. More specifically, base pairs 2, 4 and 5 of this 'recognition box' would be involved directly in this interaction (these base pairs correspond to positions 6, 4 and 3, counted from the palindromic centre to the outside of the *tet* operator in Figure 7.6), while it is speculated that base pairs 1 and 3 (positions 7 and 5 in Figure 7.6) may influence protein–DNA recognition indirectly by affecting DNA conformation.

To identify the base pairs in *tet* operator important for interaction with Tet repressor, several approaches have been taken. Three types of experiments were performed *in vitro*. Contacts to N(7) of guanine were deduced from methylation protection with dimethylsulphate (Hillen *et al.*, 1984; Klock and Hillen, 1986) and methylation interference of repressor binding after modification with dimethylsulphate. Other base contacts were scored by carbethoxylation interference of Tet repressor binding after modification of DNA with diethyl pyrocarbonate (Heuer and Hillen, 1988). Both types of methylation studies reveal two G residues, the N(7) positions of which are contacted by Tet repressor. These are at position 2 in both directions counted from the palindromic centre of O_1. The G residues at positions 6 and 8, respectively, do not yield contacts in these experiments. Also, the contacts scored by carbethoxylation interference are clearly centred around the pseudo-two-fold symmetry axis of O_1. This would indicate that the base-specific repressor–operator contacts important for *tet* operator recognition are mainly formed close to the centre of the palindrome.

Another experimental approach to identify *tet* operator–Tet repressor contacts is the saturation mutagenesis of *tet* operator. For technical reasons, this analysis was done such that one half-side of O_1 including the central base pair was mutagenized. Chemical synthesis yielded 30 mutant *tet* operators, covering all possible single base pair exchanges of an O_1

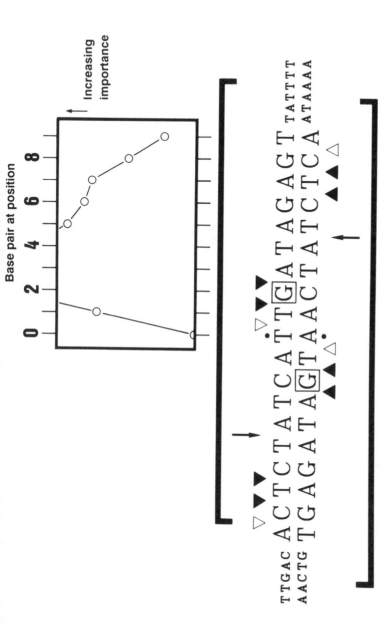

Figure 7.6 Contacts made by *tet* operator in the repressor–operator complex. The nucleotide sequence of the Tn*10*-encoded *tet* operator O₁ (large letters) is given along with flanking sequences (small letters). The dots denote the central base pair of the palindrome. The horizontal bars show the extent of DNaseI protection, with hypersensitive sites indicated by arrows. Filled triangles define essential phosphate contacts, whereas open triangles represent weaker phosphate contacts. The boxed G residues form N(7) contacts to Tet repressor. The drawing above the right *tet* operator half-side indicates the relative importance of individual positions in the palindrome for Tet repressor recognition as defined by saturation mutagenesis (see text for details). The numbers given in that figure also indicate the designation of base pairs in the *tet* operator

half-side (Wissmann *et al.*, 1988). Using an *in vivo* system, it was possible to quantitate their effects on Tet repressor binding. A sketch of the results is also depicted in Figure 7.6. While the central non-palindromic base pair does not contribute to sequence-specific recognition of Tet repressor, already the first base pair in the palindromic sequence has to be a pyrimidine for optimal affinity to the protein. This could indicate that the $N(7)$ of the purine is involved in repressor binding (Seeman *et al.*, 1976). Any sequence alteration of base pairs at the next three positions reduces Tet repressor affinity more than tenfold. Since this is the maximal reduction of affinity, which can be determined in the *in vivo* repressor titration system, this represents a minimal drop of affinity. The importance of these base pairs is greater than this loss of affinity indicates, and cannot be quantitated at present. Sequence alterations at positions five to nine of *tet* operator result in differential effects on Tet repressor binding, depending on the respective introduced base pair. The general tendency is that the quantitative importance of base pairs decreases towards the outside of the palindromic sequence of *tet* operator; however, mutations at each of the positions do affect repressor binding. It is interesting to note that all mutants except one decrease the affinity for Tet repressor. The only mutation showing increased protein binding is found at the outside end of the O_1 palindrome at position nine, where an AT base pair shows two-fold increased affinity to the repressor, compared with the wild type TA base pair (Wissmann *et al.*, 1988).

The general picture of these results agrees well with the location of modification protection and interference points on *tet* operator (see Figure 7.6). Both approaches demonstrate the importance of the inner-most base pairs at positions 2, 3 and 4, and, to a lesser extent, at positions 1 and 5 for Tet repressor recognition. Nevertheless, all base pairs of the 19 bp *tet* operator, except for the central non-palindromic one, influence Tet repressor binding as discussed above. Direct interaction with the protein has been demonstrated for base pair 2 by *in vitro* studies, and *in vivo* results indicate a contact of base pair 1 to Tet repressor. Taken together with the ethylation interference data, this would mean that base pairs 1 and 2 contribute via their phosphates and bases to repressor–operator interaction. For all other base pairs (except for position 6, which is discussed in the next chapter), it is not clear at present whether they exert their effect upon Tet repressor recognition by defining the local DNA conformation, or by directly contacting the protein. These observations have to be considered when models regarding the location of the recognition α-helix and the possible interaction of other parts of Tet repressor with *tet* operator are proposed.

Each of the *tet* operators is thought to provide two binding sites for a Tet repressor dimer. The distance of these binding sites was varied by deleting or doubling the central non-palindromic base pair of O_1. Both mutations yielded complete loss of repressor recognition in the *in vivo* assays, indicating that the two DNA binding domains of the Tet repressor dimer are not flexible enough to adjust to these structural alterations in the *tet* operator. Thus, efficient recognition requires the correct sequence and spacing of the target DNA. Taken together, the *tet* operator O_1 has, except for the ninth base pair, the optimal structure for Tet repressor binding with respect to both the nucleotide sequence and orientation of operator half sites (Wissmann *et al.*, 1988).

VII   CONTACTS OF TET REPRESSOR TO *tet* OPERATOR

Two approaches were taken to identify amino acids within the proposed operator recognition α-helix (Isackson and Bertrand, 1985) that make sequence-specific contacts to *tet* operator DNA. The first involves protein engineering of the Tet repressor to obtain mutants in which the tryptophan residues are replaced by phenylalanin (Hansen and Hillen, 1987; Hansen *et al.*, 1987). Out of the two possible exchanges in the Tn*10*-encoded Tet repressor, only the Trp to Phe exchange at position 43 in the proposed recognition α-helix reduced *tet* operator binding by about three orders of magnitude, whereas the respective change at position 75 does not affect operator binding. This indicates that Trp 43 is important for sequence-specific DNA binding. Confirmation for this assumption was obtained from the analysis of the fluorescence of Trp 43. Upon binding of *tet* operator, this Trp fluorescence is quenched to zero, while binding on non-specific DNA reduces the quantum yield only to 60 per cent (Hansen and Hillen, 1987). Thus, the thermodynamic effect is confirmed by spectroscopic results obtained in solution with the non-mutated DNA-binding motif of Tet repressor. The base pair of the *tet* operator sequence contacted by Trp 43 is not known. It is very interesting to speculate about the nature of the Trp-operator contact. The 100-per-cent quench of Trp fluorescence upon complex formation would suggest intercalation of the indole residue into the DNA (Helene and Dimicoli, 1972). It has been shown before that intercalation of Trp into double-stranded nucleic acids is a slow process. The association of the specific Tet repressor–*tet* operator complex formation is also a slow process, being roughly in the same time range (Pörschke and Ronnenberg, 1981; Kleinschmidt *et al.*, 1988). At present, there are no data excluding intercalation of Trp 43 into *tet* operator DNA upon formation of the repressor–operator complex. However, this interesting hypothesis needs to be proven.

The second approach to identify protein contacts to the operator is based upon a comparison of the class A and class B *tet* determinants (see Figures 7.2 and 7.3). The operator sequences differ at position 6 of the palindrome, where the class A operator contains an AT and the class B operator a GC base pair. The respective Tet repressor proteins exhibit strong binding to the homologous and weaker binding to the respective heterologous *tet* operator sequences (Klock *et al.*, 1985; Klock and Hillen, 1986). A comparison of the primary repressor structures reveals several differences in the region of the α-helix–turn–α-helix DNA binding motifs (see Figure 7.5). The oligonucleotide-directed exchange of Thr 40 in the class B repressor to Ala leads to a single amino acid repressor mutant, with greatly increased recognition for the class A *tet* operator carrying an AT base pair at position 6. This effect is quantitatively increased when, in addition to this amino acid, the entire recognition α-helix sequence of the class B repressor is replaced by the respective sequence of the class A repressor (see Figure 7.5). Alternatively, exchange of the entire first α-helix sequence in the α-helix–turn–α-helix motif in addition to this amino acid has the same result and changing the entire α-helix–turn–α-helix structural motif sequence reverses the recognition of both operators (Altschmied *et al.*, 1988). It is thus concluded that the nucleotide sequence specificity of Tet repressor–*tet* operator recognition for base pair 6 is determined by the chemical nature of the amino acid side chain at position 40 in the proposed recognition α-helix of the repressor sequences, and by the precise location of this α-helix in the major groove of the operator DNA, which may be determined by the entire α-helix–turn–α-helix primary structure.

Within this structural motif, the amino acids Gln 38, Pro 39 and Thr 40 in the recognition α-helix and Thr 27 and Arg 28 in the preceding α-helix have been identified to be important for *tet* operator recognition by random mutagenesis of the *tetR* gene (Isackson and Bertrand, 1985).

Since Thr 40 recognizes base pair 6 of the operator palindrome, and the phosphate contacts suggest that the sequence TCTATC interacts with the recognition α-helix of Tet repressor (Ebright, 1986), it may be speculated that this α-helix shows the opposite orientation, as suggested by homology to the λ cI and *cro* structures.

VIII   SPATIAL STRUCTURES OF TET REPRESSOR AND THE TET REPRESSOR–*tet* OPERATOR COMPLEX

The spatial structures of the Tn*10*-encoded Tet repressor dimer and its complex with a single *tet* operator, as well as with the wild type tandem *tet* operator sequence, have been determined in solution by small-angle neutron scattering (Lederer *et al.*, 1989). The repressor dimer is an elongated molecule with a maximum dimension of 10.5 ± 1.0 nm. It is best approximated by a side-by-side arrangement of two identical ellipsoids with half-axes of 3.6, 2.2 and 1.0 nm. This structure of Tet repressor is not altered to an extent noticeable in neutron scattering upon complex formation with a single *tet* operator. A DNA fragment containing the two *tet* operators O_1 and O_2 binds two Tet repressor dimers, showing a centre-to-centre distance of 11.3 ± 0.6 nm on the DNA. This is in fair agreement with the distance of 10.2 nm anticipated for the B-DNA structure of an elongated 30 bp DNA sequence (Wells *et al.*, 1980). This result would suggest that no major alteration of the DNA trajectory occurs in the *tet* transcriptional control sequence upon Tet repressor binding.

Basically, the same result has been obtained from rotational relaxation studies of the Tn*10*-encoded Tet repressor and its complex with all four wild type *tet* regulatory sequences (Pörschke *et al.*, 1988). The data of these experiments were best fitted by a repressor–operator structure involving very smooth bending of the complexed *tet* control DNA with a radius of approximately 50 nm. The same result is obtained with all four *tet* control sequences, despite the fact that the distance between the tandem *tet* operators O_1 and O_2 varies between 28 and 32 bp (Klock *et al.*, 1985). Although this independence of structure on distance does not totally exclude protein–protein contacts as the reason for smooth bending, it makes them rather unlikely. The dimensions fitting the orientation relaxation time of the native Tet repressor dimer are fitted best by a prolate ellipsoid of 10 nm for the long and 4 nm for the short axis. This is quite similar to the shape obtained from neutron scattering. Binding of the inducer tetracycline to Tet repressor does not lead to different rotation relaxation times, which indicates that no major structural changes occur in the protein upon binding of inducer.

REFERENCES

Altenbuchner, J., Schmid, K. and Schmitt, R. (1983). Tn*1721*-encoded tetracycline resistance: mapping of structural and regulatory genes mediating resistance. *J. Bacteriol.*, **153**, 116–123

Altschmied, L. and Hillen, W. (1984). Tet repressor–*tet* operator complex formation induces conformational changes in the *tet* operator DNA. *Nucl. Acids Res.*, **12**, 2171–2180

Altschmied, L., Baumeister, R., Pfleiderer, K. and Hillen, W. (1988). A threonine to alanine exchange at position 40 of Tet repressor alters the recognition of the sixth base pair of tet operator from GC to AT. *EMBO J.*, **7**, 4011–4017

Beck, C. F., Mutzel, R., Barbe, J. and Müller, W. (1982). A multifunctional gene (*tetR*) controls Tn*10*-encoded tetracycline resistance. *J. Bacteriol.*, **150**, 633–642

Berg, O. G. and von Hippel. P. H. (1985). Diffusion-controlled macromolecular interactions. *Ann. Rev. Biophys. Chem.*, **14**, 131–160

Bertrand, K. P., Postle, K., Wray, L. V., Jr and Reznikoff, W. S. (1983). Overlapping divergent promoters control expression of Tn*10* tetracycline resistance. *Gene*, **23**, 149–156

Chopra, I., Howe, T. G. B., Linton, A. H., Linton, K. B., Richmond, M. H. and Spelber, D. C. E. (1981). The tetracyclines: prospects at the beginning of the 1980s. *J. Antimicrob. Chemother.*, **5**, 5–21

Ebright, R. H. (1986). Proposed amino acid–base pair contacts for 13 sequence-specific DNA binding proteins. *Protein Structure, Folding, and Design.* In UCLA Symposia on Molecular and Cellular Biology, Vol. 39 (Oxender, D. L., ed.), Liss, New York, 207–219

Epe, B. and Woolley, P. (1984). The binding of 6-demethylchlortetracycline to 70S, 50S, and 30S ribosome particles: a quantitative study by fluorescence anisotropy. *EMBO J.*, **3**, 119–126

Hansen, D., Altschmied, L. and Hillen, W. (1987). Engineered Tet repressor mutants with single tryptophan residues as fluorescent probes. *J. Biol. Chem.*, **262**, 14030–14035

Hansen, D. and Hillen, W. (1987). Tryptophan in α-helix 3 of Tet repressor forms a sequence-specific contact with *tet* operator in solution. *J. Biol. Chem.*, **262**, 12269–12275

Helene, C. and Dimicoli, J. (1972). Interaction of oligopeptides containing aromatic amino acids with nucleic acids. Fluorescence and proton magnetic resonance studies. *FEBS Lett.*, **26**, 6–10

Hershfield, V., Boyer, H. W., Chow, L. and Helinski, D. R. (1976). Characterization of a mini-ColE1 plasmid. *J. Bacteriol.*, **126**, 447–453

Hever, C. and Hillen, W. (1988). Tet repressor–tet operator contacts probed by operator DNA-modification interference studies. *J. Mol. Biol.*, **202**, 407–415

Hillen, W., Gatz, C., Altschmied, L., Schollmeier, K. and Meier, I. (1983). Control of expression of the Tn*10*-encoded tetracycline resistance genes: equilibrium and kinetic investigation of the regulatory reactions. *J. Mol. Biol.*, **169**, 707–722

Hillen, W., Klock, G., Kaffenberger, I., Wray, L. V., Jr and Reznikoff, W. S. (1982a). Purification of the *Tet* repressor and *tet* operator from the transposon Tn*10* and characterization of their interaction. *J. Biol. Chem.*, **257**, 6605–6613

Hillen, W., Klein, R. B. and Wells, R. D. (1981). Preparation of milligram amounts of 21 deoxyribonucleic acid restriction fragments. *Biochemistry*, **20**, 3748–3756

Hillen, W. and Schollmeier, K. (1983). Nucleotide sequence of the Tn*10* encoded tetracycline resistance gene. *Nucl. Acids Res.*, **11**, 525–539

Hillen, W., Schollmeier, K. and Gatz, C. (1984). Control of expression of the Tn*10*-encoded tetracycline resistance gene. II. Interaction of RNA polymerase and Tet repressor with the *tet* operon regulatory region. *J. Mol. Biol.*, **172**, 185–201

Hillen, W., Unger, B. and Klock, G. (1982b). Analysis of *tet* operator-Tet repressor complexes by thermal denaturation studies. *Nucl. Acids Res.*, **10**, 6085–6097

Hillen, W. and Unger, B. (1982). Binding of four repressors to double-stranded *tet* operator region stabilizes it against thermal denaturation. *Nature*, **297**, 700–702

Hochschild, A. and Ptashne, M. (1986). Cooperative binding of λ repressors to sites separated by integral turns of the DNA helix. *Cell*, **44**, 681–687

Isackson, P. J. and Bertrand, K. P. (1985). Dominant negative mutations in the Tn*10* Tet repressor. Evidence for use of the conserved helix–turn–helix motif in DNA binding. *Proc. Natl Acad. Sci. USA*, **82**, 6226–6230

Izaki, K., Kiuchi, K. and Arima, K. (1966). Specificity and mechanism of tetracycline resistance in a multiple drug resistant strain of *Escherichia coli*. *J. Bacteriol.*, **91**, 628–633

Jorgensen, R. A. and Reznikoff, W. S. (1979). Organization of structural and regulatory genes that mediate tetracycline resistance in transposon Tn10. *J. Bacteriol.*, **138**, 705–714

Kleinschmidt, C., Tovar, K., Hillen, W. and Pörschke, D. (1988). Dynamics of a Repressor–Operator Recognition: the Tn*10* encoded tetracycline resistance control. *Biochemistry*, **27**, 1094–1104

Klock, G. and Hillen, W. (1986). Expression, purification and operator binding of the transposon Tn*1721*-encoded Tet repressor. *J. Mol. Biol.*, **189**, 633–641

Klock, G., Unger, B., Gatz, C., Hillen, W., Altenbuchner, J., Schmid, K. and Schmitt, R. (1985). Heterologous repressor–operator recognition among four classes of tetracycline resistance determinants. *J. Bacteriol.*, **161**, 326–332

Lederer, H. Tovar, K., Baer, G., May, R. P., Hillen, W. and Heumann, H. (1989). The quaternary structure of Tet repressors bound to the Tn*10* encoded *tet* gene control region determined by neutron solution scattering. *EMBO J.*, in the press

Levy, S. B. and McMurray, L. (1978). Plasmid-determined tetracycline resistance involves new transport systems for tetracycline. *Nature*, **275**, 90–92

Levy, S. B. (1988). Tetracycline resistance determinants are widespread. *ASM News*, **54**, 418–421

Little, J. W., Edmiston, S. H., Pacelli, L. Z. and Mount, D. W. (1980). Cleavage of the *Escherichia coli lexA* protein by the *recA* protease. *Proc. Natl Acad. Sci. USA*, **77**, 3225–3229

Liu-Johnson, H-N., Gartenberg, M. R. and Crothers, D. M. (1986). The DNA bending domain and bending angle of *E. coli* CAP protein. *Cell*, **47**, 681–687

McMurray, L., Petrucci, R. E. and Levy, S. B. (1980). Active efflux of tetracycline encoded by four genetically different resistance determinants in *Escherichia coli*. *Proc. Natl Acad. Sci. USA*, **77**, 3974–3977

Marshall, B., Morrisey, S., Flynn, P. and Levy, S. B. (1986). A new tetracycline resistance determinant, class E, Isolated from *Enterobacteriaceae*. *Gene*, **50**, 111–117

Meier, I., Wray, L. V., Jr and Hillen, W. (1988). Differential regulation of the Tn*10* encoded tetracycline resistance genes *tetA* and *tetR* by the tandem *tet* operators O_1 and O_2. *EMBO J.*, **7**, 567–572

Mendez, B., Tachibana, C. and Levy, S. B. (1980). Heterogeneity of tetracycline resistance determinants. *Plasmid*, **3**, 99–108

Nguyen, T. T., Postle, K. and Bertrand, K. P. (1983). Sequence homology between the tetracycline resistance determinants of Tn*10* and pBR322. *Gene*, **25**, 83–92

Oemichen, R., Klock, G., Altschmied, L. and Hillen, W. (1984). Construction of an *E. coli* strain overproducing the Tn*10*-encoded TET repressor and its use for large scale purification. *EMBO J.*, **3**, 539–543

Ohlendorf, D. H., Anderson, W. F., Lewis, M., Pabo, C. O. and Matthews, B. W. (1983). Comparison of the structures of Cro and λ repressor proteins from bacteriophage λ. *J. Mol. Biol.*, **169**, 757–769

Pabo, C. O. and Sauer, R. T. (1984). Protein–DNA recognition. *Ann. Rev. Biochem.*, **53**, 293–321

Pörschke, D. and Ronnenberg, J. (1981). The reaction of aromatic peptides with double helical DNA. Quantitative characterization of a two step reaction scheme. *Biophys. Chem.*, **13**, 283–290

Pörschke, D., Tovar, K. and Antosiewicz, J. (1988). Structure of the Tet repressor–operator complexes in solution from electrooptical measurements and hydrodynamic simulations. *Biochemistry*, **27**, 4674–4679

Postle, K., Nguyen, T. T. and Bertrand, K. P. (1984). Nucleotide sequence of the repressor gene of the Tn*10* tetracycline resistance determinant. *Nucl. Acids Res.*, **12**, 4849–4863

Ptashne, M., Backman, K., Humayun, M. Z., Jeffrey, A., Maurer, R., Meyer, B. and Sauer, R. T. (1976). Autoregulation and function of a repressor in bacteriophage lambda. *Science*, **194**, 156–161

Ptashne, M., Jeffrey, A., Johnson, A. D., Maurer, R., Meyer, B. J., Pabo, C. O., Roberts, T. M. and Sauer, R. T. (1980). How the λ repressor and cro work. *Cell*, **19**, 1–11

Roberts, J. W. and Roberts, C. W. (1975). Proteolytic cleavage of bacteriophage lambda repressor in induction. *Proc. Natl Acad. Sci. USA*, **72**, 147–159

Sauer, R. T., Yocum, R. R., Doolittle, R. F., Lewis, M. and Pabo, C. O. (1982). Homology among DNA-binding proteins suggests use of a conserved super-secondary structure. *Nature*, **298**, 447–451

Schevitz, R. W., Otwinowski, Z., Joachimiak, A., Lawson, C. L. and Sigler, P. B. (1985). The three-dimensional structure of *trp* repressor. *Nature*, **317**, 782–786

Seeman, N. C., Rosenberg, J. M. and Rich, A. (1976). Sequence-specific recognition of double helical nucleic acids by proteins. *Proc. Natl Acad. Sci. USA*, **73**, 804–808

Takahashi, M., Altschmied, L. and Hillen, W. (1986). Kinetic and equilibrium characterization of the Tet repressor–tetracycline complex by fluorescence measurements. *J. Mol. Biol.*, **107**, 341–348

Tovar, K., Ernst, A. and Hillen, W. (1988). Identification and nucleotide sequence of the class E tet regulatory elements and operator and inducer binding of the encoded purified Tet repressor. *Mol. Gen. Genet.*, **215**, 76–80

Unger, B., Becker, J. and Hillen, W. (1984a). Nucleotide sequence of the gene, protein purification and characterization of the pSC101-encoded tetracycline resistance-gene-repressor. *Gene*, **31**, 103–108

Unger, B., Klock, G. and Hillen, W. (1984b). Nucleotide sequence of the repressor gene of the RA1 tetracycline resistance determinant: structural and functional comparison with three related Tet repressor genes. *Nucl. Acids Res.*, **12**, 7693–7703

Wagenhöfer, M., Hansen, D. and Hillen, W. (1988). Thermal denaturation of engineered Tet repressor proteins and their complexes with tet operator and tetracycline studied by temperature gradient gel electrophoresis. *Ann. Biochem.*, **175**, 422–432

Waters, S., Rogowsky, P., Grinsted, J., Altenbuchner, J. and Schmitt, R. (1983). The tetracycline resistance determinants of RP1 and Tn*1721*: nucleotide sequence analysis. *Nucl. Acids Res.*, **11**, 6089–6105

Weber, K. and Geisler, N. (1978). *lac* repressor fragments produced *in vivo* and *in vitro*: an approach to the understanding of the interaction of repressor and DNA. In *The Operon*, Reznikoff, W. S. and Miller, J. (eds), Cold Spring Harbour Laboratory Press, New York, 155–175

Wells, R. D., Goodman, T. C., Hillen, W., Horn, G. T., Klein, R. D., Larson, J. E., Müller, U. R., Neuendorf, S. K., Panayotatos, N. and Stirdivant, S. M. (1980). *Prog. Nucl. Acids Res. Molec. Biol.*, **24**, 167–267

Wissmann, A., Meier, I. and Hillen, W. (1988). Saturation mutagenesis of the Tn*10*-encoded *tet* operator O_1: identification of base pairs involved in Tet repressor recognition. *J. Mol. Biol.*, **202**, 397–406

Wray, L. V., Jr, Jorgensen, R. A. and Reznikoff, W. S. (1981). Identification of the tetracycline resistance promoter and repressor in transposon Tn*10*. *J. Bacteriol.*, **147**, 297–304

Wray, L. V., Jr and Reznikoff, W. S. (1983). Identification of repressor binding sites controlling expression of tetracycline resistance encoded by Tn*10*. *J. Bacteriol.*, **156**, 1188–1191

Zhang, R-G., Joachimiak, A., Lawson, C. L., Schevitz, R. W., Otwinowski, Z. and Sigler, P. B. (1987). The crystal structure of *trp* aporepressor at 1.8 Å shows how binding tryptophan enhances DNA affinity. *Nature*, **327**, 591–597

8
Structure and condensation of chromatin

M. H. J. Koch

INTRODUCTION

Various aspects of chromatin structure and function have recently been reviewed (Igo-Kemenes *et al.*, 1982; Annunziato and Seale, 1983; Eissenberg *et al.*, 1985; Felsenfeld and McGhee, 1986; Nagl, 1986; Nelson *et al.*, 1986; Pederson *et al.*, 1986; Wu *et al.*, 1986; Yaniv and Cereghini, 1986; Sayers, 1988). The present chapter concentrates on the higher-order organization of the chromatin fibre and on some of the physico-chemical factors controlling this organization. It is not intended as a review, and the references merely provide convenient entry points to the maze of the literature on chromatin. We have, however, tried to present the competing models and indicate where there appear to be contradictions or inconsistencies. If, at the end, a number of questions are clarified – as opposed to problems solved – then our goal will have been reached.

THE CHROMATIN STRUCTURE

Packing of long DNA molecules in a small volume is a recurrent feature in biological structures. In the nucleus of eukaryotic organisms, this is achieved by chromatin, the regular complex of DNA with basic histone proteins. Formation of this complex drastically reduces the apparent length of DNA through a hierarchy of folded structures, yet allows the control and transcription processes to take place. In the metaphase nucleus, the genome is partitioned into chromosomes that are about 1400 nm wide and 5 μm long, and each contains about 5 cm of DNA. They are formed of condensed sections about 700 nm wide, which themselves consist of looped domains that are about 400 nm wide. These domains result from the folding of the 30 nm filament described by Finch and Klug (1976). Fragments of the 30 nm filament can be excised by nuclease digestion and reversibly unfolded at low ionic strength. The regular beaded structure, which gives rise to the characteristic ladder in DNA electrophoresis gels, then becomes clearly visible in electron micrographs (Thoma *et al.*, 1979). The beads are the nucleosome core particles that can be obtained by prolonged nuclease digestion. They are flat disks, with a diameter of 11 nm and a height of 5.7 nm. The crystal structure of these core particles has been determined by Richmond *et al.* (1984) to 0.7 nm resolution. The particle consists of 1.8 turns of a left-handed superhelix of B-DNA wrapped around an octamer of the histones. The latter is made up of a tetramer of H3–H4 and of two H2A–H2B dimers. The DNA superhelix has an average pitch of 2.7 nm. The protein does not extend outside the superhelix, nor between the turns. The chromatosome is similar to the nucleosome, but has an additional 20 base pairs of DNA, completing two turns of the DNA superhelix, together with the H1 histone. Successive chromatosomes are connected by a stretch of DNA – the linker DNA – consisting of a nearly constant number of base pairs (Prunell and Kornberg, 1982). The number of base pairs in the linker depends, however, on the source of the chromatin. It varies from about 15 base pairs in neuronal chromatin to about 50 base pairs in chicken erythrocyte chromatin and 80 base pairs in sea urchin sperm chromatin. The chromatosome and the linker DNA together form the actual monomer or repeating unit of the chromatin fibre. Figure 8.1 illustrates that the various levels of folding and the ensuing reduction of apparent length can be described as a fractal structure (Mandelbrot, 1983), with a Hausdorff dimension of 2.3. Although these general features apply to chromatin from various sources, there is still a large variety in the histones, the separation of the nucleosomes along the DNA, the diameter of the condensed fibre, the details of the structure of the chromosomes and their number.

THE STRUCTURE OF CHROMATIN IN SOLUTION

For the purpose of modelling the low resolution structure for comparison with X-ray or neutron solution scattering data (for an introduction to X-ray solution scattering, see Porod, 1982; for neutron scattering on chromatin, see Bradbury and Baldwin, 1986), the simple model of the chromatosome in Figure 8.2A suffices. In fact, the scattering is dominated by the core DNA. The fractions of the excess electrons in the core DNA, the linker DNA and the histone octamer are 0.51, 0.18 and 0.31 respectively. This allows simplification of the calculation of the scattering patterns even further, by taking into account only the contribution of the core DNA as shown in Figure 8.2B. The scattering patterns calculated for these models in Figure 8.2C display bands at values of s ($s = 2 \sin\theta/\lambda$, where 2θ is the scattering angle and λ the wavelength) around $(5.7\ \text{nm})^{-1}$, $(3.7\ \text{nm})^{-1}$ and $(2.7\ \text{nm})^{-1}$, as well as the local minimum at $(4.5\ \text{nm})^{-1}$. Such values are in

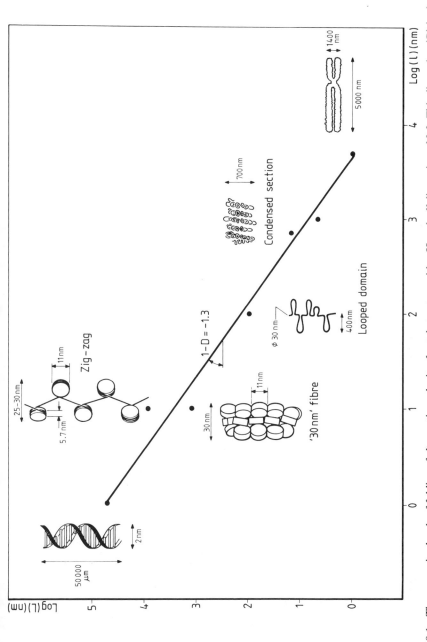

Figure 8.1 The successive levels of folding of chromatin make it a fractal structure with a Hausdorff dimension of 2.3. This dimension (*D*) is obtained from the slope of the plot of the logarithm of the apparent length against the logarithm of the length of the ruler

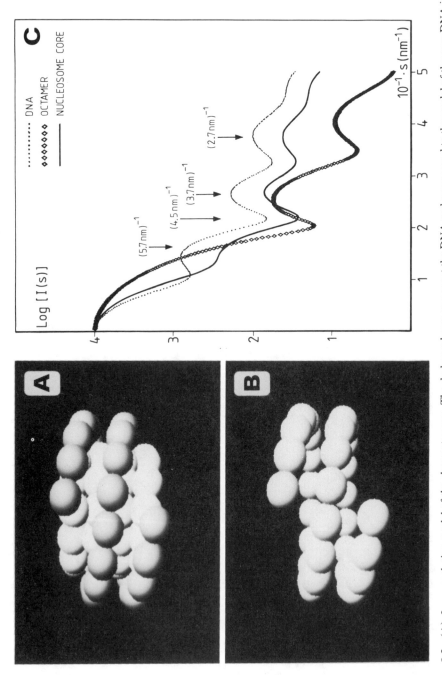

Figure 8.2   (A): Low-resolution model of the chromatosome. The darker spheres represent the DNA and correspond to the model of the core DNA in (B) used in the calculation of the X-ray solution scattering patterns. (C): Calculated patterns of the chromatosome, the DNA and the octamer model

agreement with the experimental patterns obtained by Damaschun *et al.* (1980).

As the contribution of the linker is negligible at low resolution, the solution scattering pattern of chromatin should mainly reflect the relative arrangement of the nucleosomes, as well as their intrinsic features. Since the largest dimension of the nucleosomes is 11 nm, their scattering pattern does not present any features, apart from a smooth decay, for *s*-values below $(11 \text{ nm})^{-1}$. This divides the scattering pattern of chromatin in two regions, as illustrated in Figure 8.3. The region below $s = (11 \text{ nm})^{-1}$ can only reflect changes in the higher-order structure, and also interfibre interactions in the case of concentrated solutions. The region at higher *s*-values is mainly sensitive to the internal structure of the nucleosomes and to the details of their short-range packing in condensed chromatin. Note that the characteristic features of the calculated nucleosome pattern are all present in the experimental solution patterns of chromatin. These bands are detected in solution, as well as in the patterns of oriented gels, as illustrated in Figure 8.3C. The latter give unequivocal evidence for the preferred orientation of the nucleosomes in the fibres. The broad arcs are isolated interference bands characteristic of a system with short-range order, rather than reflections originating from diffraction by a lattice.

As long as there is no aggregation, the solution scattering pattern reflects the features of the isolated fibres. In concentrated solutions, gels, oriented fibres and nuclei, interference bands due to interactions between fibres can appear. Studies on dilute and on concentrated systems are thus both required to make unequivocal assignments of the scattering bands. The features of the pattern for $s > 0.1 \text{ nm}^{-1}$ do not change much with the degree of condensation, indicating that condensation does not affect the structure of the nucleosomes. This contrasts with the drastic changes upon condensation at *s*-values below 0.1 nm^{-1}. Three specific features of the patterns in Figure 8.3 play a role in the further interpretation. In the pattern of solutions and gels at very low ionic strength, there is a strong interference band near $(20 \text{ nm})^{-1}$ that shifts to higher *s*-values and vanishes when the ionic strength is increased. This band in the pattern of uncondensed chromatin has been assigned to an interference due to the regular spacing of the nucleosomes (Perez-Grau *et al.*, 1984; Bordas *et al.*, 1986a, b; Koch *et al.*, 1987a, b). Although it has not always been interpreted as such, it has been observed in all recent scattering studies of chromatin at very low ionic strength (for a survey see Koch *et al.*, 1987a). The band is also clearly visible in the patterns obtained by Widom (1986) and Gerchman and Ramakrishnan (1987).

In the pattern of condensed chromatin in Figure 8.3B, a weak band corresponding to the first side maximum of the transform of the fibre appears in the range $(20\text{--}30 \text{ nm})^{-1}$. This is the band also observed by Williams *et al.* (1986) in the patterns of nuclei. In the patterns of gels or

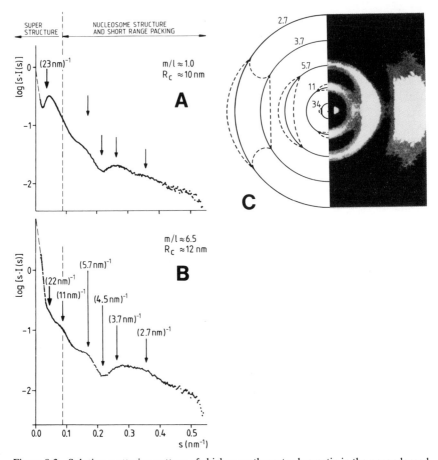

Figure 8.3   Solution scattering pattern of chicken erythrocyte chromatin in the uncondensed state (A) and in the condensed state (B). The positions of features characteristic of the nucleosome structure and of the bands resulting from the superstructure are indicated (after Bordas *et al.*, 1986a). (C): Small-angle diffraction pattern of an oriented gel of condensed chromatin illustrating the meridional or equatorial origin of the interference bands (after Widom and Klug (1985))

nuclei, there is often also a band in the range $(30\text{–}40 \text{ nm})^{-1}$, resulting from the side-by-side packing of fibres. It appears, as expected, on the equator of the patterns of oriented fibres (Widom and Klug, 1985), as illustrated in Figure 8.3C.

It is useful to consider the arguments leading to the assignment of these bands and the verifications that were made.

UNCONDENSED CHROMATIN

The primary evidence for the regular spacing of the nucleosomes comes from electron microscopic observations (Olins and Olins, 1974; Oudet *et al.*, 1975; Thoma *et al.*, 1979). The images of the zig-zag structure in chicken erythrocyte chromatin (Thoma *et al.*, 1979) suggest that there is a regular internucleosomal distance of about 20 nm, which in turn implies that the linker DNA is nearly straight.

A structure like the zig-zag can be described as the convolution of a nucleosome with a chain. With minor modifications, this is also how the structures are generated for the computer simulations, as illustrated in Figure 8.4. Since the distance between the nucleosomes is about twice their diameter, their relative orientation can be neglected in first approximation, and they can be represented by their spherical average. The variables defining the chain are then only the distance d between successive nucleosomes, the angle θ and the dihedral angle ϕ defined in Figure 8.4. The intensity scattered by such an object is given by Debye's formula (see, for instance, Cantor and Schimmel, 1980):

$$I(s) = f_N^2(s) \cdot \sum_i \sum_j \frac{\sin(2\pi s r_{ij})}{2\pi s r_{ij}} \tag{8.1}$$

The intensity corresponding to the scattering vector s is thus the product of the scattering of the spherically averaged nucleosome (f_N^2) and that of the chain determined solely by the distances r_{ij} between the ith and jth nucleosomes.

An analysis of the contribution of the various terms in equation (8.1) to this pattern gives indications about the origin of the interference band near $(20 \text{ nm})^{-1}$. For this purpose, it is useful to represent the double sum in this equation as the sum of the contribution of all distances between ith and jth nucleosomes as a function of s, as illustrated in Figure 8.5. This suggests that the main contribution to the band near $(20 \text{ nm})^{-1}$ is due to the distance between nearest neighbouring nucleosomes. There are several ways to verify this.

The position of this band should, other things being equal, depend on the length of the linker DNA, and should also affect the value of the radius of gyration of the cross-section. Scattering experiments on sea urchin sperm chromatin, which has a repeat of about 245 base pairs, as compared with 210 base pairs in chicken erythrocyte chromatin confirm this, as illustrated in Figure 8.6. No reliable scattering data are yet available for chromatin with very short linker, but the model predicts that for neuronal chromatin, which has a repeat of 165 base pairs, the interference maximum should be shifted to s-values around $(15 \text{ nm})^{-1}$, and should thus also be less intense. The radius of gyration of the cross-section should also be significantly lower than in chicken erythrocyte chromatin, about 6 nm instead of 10 nm.

Another verification of the assignment of the $(20 \text{ nm})^{-1}$ band comes

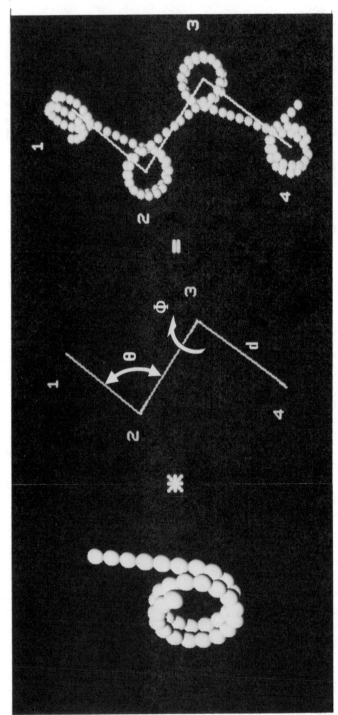

Figure 8.4   Approximations to the structure of chromatin at low ionic strength. Convolution of the model of a chromatosome and linker DNA with a chain with 23 nm segment length. d is the centre-to-centre distance between nucleosomes, θ the angle between adjacent segments and ϕ the dihedral angle between the planes defined by three successive segments

from nuclease digestion experiments (Koch *et al.*, 1987b). For a given *s*-value, the double sum in equation (8.1) can be represented by a symmetric matrix as illustrated in Figure 8.7. When the nucleosome chain is cut, all dashed terms in the top right-hand side and bottom left-hand side matrices vanish. As illustrated in Figure 8.5, these terms, for which $i-j>1$, are small. The main change in the scattering pattern will thus result from the disappearance of one of the nearest neighbour terms, indicated by the unmarked white squares. Note that the contribution of the diagonal terms remains unchanged. Obviously, after one cut, the remaining scattering is represented by the two smaller symmetric matrices corresponding to the two fragments of the chain. To interpret the evolution of the scattering pattern during nuclease digestion illustrated in Figure 8.8A, it is useful to rewrite equation (8.1) for a polydisperse solution. The contributions to the scattering of the terms indicated by the roman numerals in Figure 8.7 then become apparent:

$$I(s,t) = f^2 \sum_{k=1}^{K} k\, N_k(t) \qquad\qquad \dots \text{I}$$

$$+ 2f^2 \frac{\sin(2\pi s d)}{2\pi s d} \sum_{k=2}^{K} (k-1)\, N_k(t) \qquad \dots \text{II} \qquad (8.2)$$

$$+ 2f^2 \sum_{k=3}^{K} \sum_{m=2}^{k-1} (k-m)\, N_k(t)\, \frac{\sin(2\pi s m d)}{2\pi s m d} \dots \text{III}$$

In this equation, k is the degree of polymerization, which can take values between 1 and K, and $N_k(t)$ is the number of k-mers in solution. $\sum_{k=1}^{K} k N_k(t)$ is the total number of nucleosomes in solution which remains constant, and $\sum_{k=2}^{K} (k-1)N_k(t)$ is the total but decreasing number of links between nucleosomes. The first term thus corresponds to the diagonal of the matrix (i.e. the constant term in Figure 8.5), the second one to the nearest neighbour interactions and the last one to all remaining off-diagonal elements. The nearest neighbour interference function can be obtained by dividing the scattering pattern at time *t* by that of the mononucleosomes. In good approximation, the latter corresponds to the pattern obtained after prolonged digestion. The resulting interference function in Figure 8.8B is in excellent agreement with the theoretical curve for an average internucleosomal distance of 23 nm. The amplitude of the maxima of this function is proportional to the number of internucleosomal links. The decrease of this amplitude as a function of time in Figure 8.8C agrees with the standard expressions of first-order kinetics polymer degradation, as expected for random cutting of the chain by micrococcal nuclease.

Having established the origin of the interference band, it is possible to use it as a sensitive marker of the structural changes in the chromatin fibre at low ionic strength. The effect of temperature, intercalating dyes like

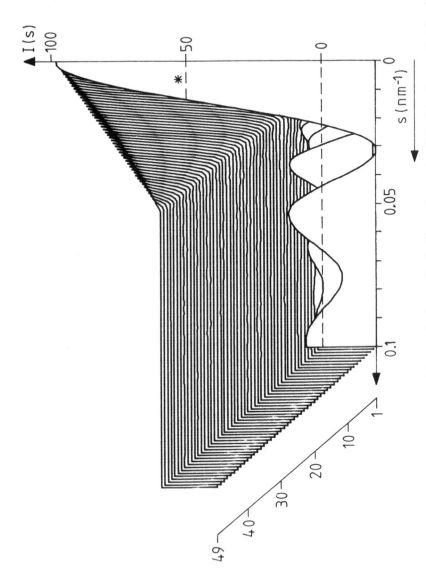

Figure 8.5 Plot as a function of the scattering vector s of the contributions to the double sum in equation (8.1) of the distances $i - j$ ($i = j$) between ith and jth nucleosomes for a chain of 50 nucleosomes with an average separation of 23 nm. The average of the scattering patterns of 100 conformations with a gaussian distribution of θ around 110 degrees and a uniform distribution of ϕ was taken. The level of the constant term ($i = j$) is indicated by the asterisk

ethidium bromide or binding of netropsin and distamycin were investigated. In all cases, the observations could be explained at least qualitatively by the model described earlier (Bordas *et al.*, 1986b; Koch *et al.*, 1987a, b).

Radius of gyration of uncondensed chromatin

The values of the radius of gyration obtained by light scattering (Ausio *et al.*, 1984) on fractionated chromatin provide further evidence for the type of model illustrated in Figure 8.4. For a chain with *n* identical segments of length *d*, fixed angles θ and without restrictions on the torsion angles ϕ, the radius of gyration is $R_g = [\frac{1}{6}nd^2(1 - \cos\theta)/(1 + \cos\theta)]^{1/2}$ (see, for instance, Flory, 1953). Table 8.1 gives a comparison between the experimental and theoretical values. Note that Ausio *et al.* (1983) had already used a similar model suggested by E. Trifonov to obtain a good agreement with the experimental data.

Other hydrodynamic results

There have been several attempts to calculate the frictional properties of uncondensed chromatin (Ramsay-Shaw and Schmitz, 1976; Schmitz and Ramsay-Shaw, 1977; Fulmer and Bloomfield, 1982). The conclusion of these studies has been that uncondensed chromatin in solution cannot adequately be represented by a 10 nm filament. Modelling devices such as helices and coils were introduced to fit the experimental results. Calculations using the type of model illustrated in Figure 8.4 have not yet been carried out.

CONDENSED CHROMATIN

The fundamental structural parameters, mass per unit length and fibre diameter, for chicken erythrocyte chromatin have been obtained by different methods. The values of the radius of gyration measured by light scattering (Ausio *et al.*, 1984) or X-ray scattering (Lasters *et al.*, 1985) are compatible with a fibre diameter of 30 nm. Neutron and X-ray scattering give a radius of gyration of the cross-section corresponding also to a fibre diameter around 30 nm, and a mass per unit length around 6.5 nucleosomes/11 nm (Suau *et al.*, 1979; Bordas *et al.*, 1986a; Greulich *et al.*, 1987; Koch *et al.*, 1987a; Gerchman and Ramakrishnan, 1987). All these observations have thus fully confirmed the values originally given by Finch and Klug (1976).

Values of the mass/unit length up to 12 nucleosomes/11 nm have been obtained by electron microscopy for chicken erythrocyte chromatin at physiological salt concentrations (Woodcock *et al.*, 1984). A similar value

Table 8.1 Radius of gyration (R_g) of chicken erythrocyte chromatin determined by light scattering (Ausio *et al.*, 1984) for fractions with different weight average (N_w) and z, $z+1$ average ($N_{z,z+1}$) number of nucleosomes per chromatin chain (for definitions of these averages, see e.g. Tanford, 1961). $\theta_{z,z+1}$ and θ_w are the angles (in degrees) between segments of 23 nm length corresponding to the R_g values, assuming an unrestricted chain.

[NaCl] (mM)	R_g (nm)	$N_{z,z+1}$	$\theta_{z,z+1}$	N_w	θ_w
5	42.6	24.5	86	18	95
5	74.3	48	98	35	107
5	81.7	58	98	42	107
1	102.2	58	110	42	119

was inferred from sedimentation results for rat liver chromatin (Walker and Sikorska, 1987). Although there is evidence that with most solution techniques, full compaction may not be reached before precipitation sets in, these values appear too large because they imply packing ratios equal or higher than in dense crystalline solids.

Reports indicating that the fibre diameter is independent of the linker length (McGhee *et al.*, 1983; Thomas *et al.*, 1986; Widom *et al.*, 1985) have not been corroborated. On the contrary, X-ray scattering confirmed the observations of Zentgraf and Franke (1984) and Williams *et al.* (1986). The radius of gyration of the cross-section of sea urchin sperm chromatin, which has a 75 base pairs linker is significantly larger (15 nm) than in chicken erythrocyte chromatin (11 nm) (Koch *et al.*, 1988). The mass per unit length of fully condensed sea urchin sperm is also higher than in chicken erythrocyte chromatin (Williams *et al.*, 1986; Koch *et al.*, 1988).

The scattering pattern of oriented gels in Figure 8.3C (Widom and Klug, 1985) leaves no doubt that the nucleosomes are preferentially oriented, with the normal to their faces perpendicular to the fibre axis. The arcing of the interference bands, however, proves that there is a broad distribution of orientation. As a result of this, X-ray scattering patterns can be calculated by a procedure very similar to the one used for uncondensed chromatin. A density of randomly distributed points, with a constraint on the minimum distance between points, corresponding to the appropriate mass/unit length, is generated on a cylindrical surface as illustrated in Figure 8.9. The scattering of this distribution is then multiplied by that of an isolated nucleosome. The results of this simple calculation in Figure 8.10 should be compared with the experimental data in Figure 8.3B. They compare favourably with the result obtained with the more sophisticated models in Figure 8.9E. For comparison, the calculated scattering pattern obtained from the electron microscopic results of Subirana *et al.* (1985) on partially condensed chromatin is also shown. The

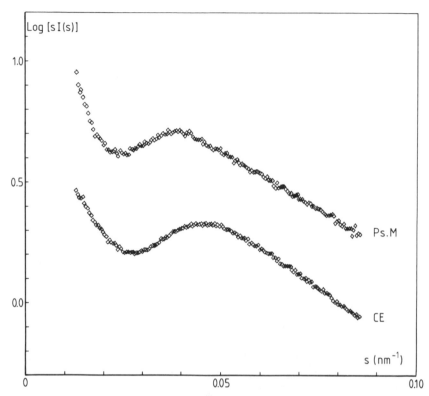

Figure 8.6   X-ray scattering patterns of sea urchin (*Psammechinus miliaris*) sperm chromatin (top) and chicken erythrocyte chromatin (bottom), illustrating the shift of the interference band to lower *s*-values with increasing linker length

calculated radius of gyration of the cross-section is too small (7 nm), probably as result of shrinking of the fibre.

It is clear that the major differences between the pattern of condensed chromatin and that of nucleosomes occur at *s*-values below $(11 \text{ nm})^{-1}$. The band near (23 nm) in the pattern of condensed chicken erythrocyte chromatin is of special interest, as it plays a role in the interpretations of Bordas *et al*. (1986a) and Williams *et al*. (1986). The band arises from the first side maximum on the equator of the fibre transform. As illustrated in the calculated patterns in Figure 8.10, and in agreement with the observations of Williams *et al*. (1986) and Koch *et al*. (1988), the band is shifted to lower *s*-values and increases in intensity when the fibre diameter increases.

The condensation properties of chromatin are largely programmed in the dinucleosomes, as shown by Perez-Grau *et al*. (1982) in an electron microscopic study on the assembly products of chicken erythrocyte mono- and dinucleosomes. In the presence of divalent cations, dinucleosomes form cylindrical aggregates with a diameter around 30 nm that closely

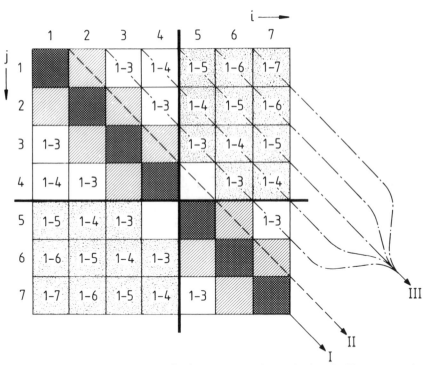

Figure 8.7   The scattering of a chain (top) at a given *s*-value can be described by a symmetric matrix representing the contribution of each of the *i*, *j* distances in the chain. When the chain is cut, all cross-terms between nucleosomes in the two parts vanish as described in equation (8.2). The contribution of the diagonal terms remains constant

resemble condensed chromatin. This suggests that hydrophobic interactions play an important role in the final packing, whereas control of compaction is mainly electrostatic and entropic, as discussed below.

CHROMATIN IN GELS AND NUCLEI

The X-ray scattering patterns of very concentrated samples of chromatin often display a band near $(30 \text{ nm})^{-1}$. This band has been observed in the pattern of nuclei (Baudy and Bram, 1979; Langmore and Schutt, 1980; Langmore and Paulson, 1983; Bordas *et al.*, 1986a; Williams *et al.*, 1986; Notbohm, 1986a) and unoriented gels (Bordas *et al.*, 1986a), as well as in oriented gels (Widom and Klug, 1985), where its intensity is concentrated on the equator. In gels, the position of the band can be shifted by altering the chromatin or salt concentrations. It cannot be made to move to *s*-values above $(30.3 \text{ nm})^{-1}$, except by removal of the H1/H5 histones, in which case the band appears at $(16.7 \text{ nm})^{-1}$ (Bordas *et al.*, 1986a). These observations suggest that the band results from the side-by-side packing of fibres with statistical rotation and shifts. In this case, interference effects occur only on the equator (Oster and Riley, 1952; Vainshtein, 1966) and the intensity is given by

$$I(s) = F^2(s)T(s) \tag{8.3}$$

where $F(s)$ is the transform of the isolated fibre and $T(s)$ the interference function describing the short-range order. The exact mode of packing of the fibres, including the average number of neighbours, has little influence on the qualitative conclusions. The interference that modulates the pattern of the isolated fibre depends on the zero-order Bessel function $J_0(2\pi as)$, where a is the centre-to-centre distance between fibres. This function has maxima that do not correspond to Bragg spacings, but occur at values of $s = (h + \frac{1}{8})/a$, where h is the integer order of the maximum. When a decreases, the maximum shifts to higher *s*-values and its intensity decreases. If the fibres are very tightly packed, the separation corresponds to their diameter, and the maximum can no longer be detected due to lack of contrast. This is possibly the case in the pattern of oriented sea urchin chromatin gels (Widom *et al.*, 1985).

The fact that in the experimental patterns only one maximum is clearly visible arises from the very low degree of order, but probably also from overlap with other features of the equatorial pattern in Figure 8.3C.

ORIENTATION OF THE NUCLEOSOMES AND OF THE LINKER DNA

Interpretation of the previous results does not depend on the orientations of the nucleosomes relative to each other, nor on the path of the linker DNA. These cannot easily, if at all, be determined by solution scattering or

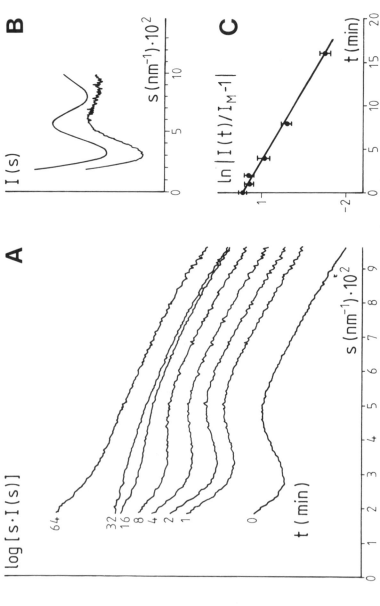

Figure 8.8 (A) X-ray solution scattering patterns of chicken erythrocyte chromatin ($A_{260} = 60$), during digestion by micrococcal nuclease. (B) Interference function obtained from the ratio of the patterns before and after 32 min digestion; top: theoretical interference function for an internucleosomal spacing of 23 nm. (C) Decrease in the number of internucleosomal links, assuming first-order kinetics

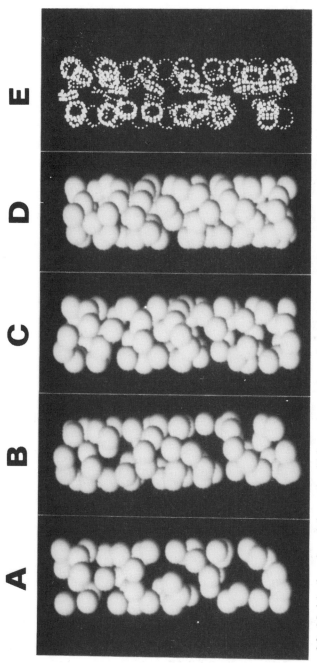

Figure 8.9 Models of condensed chromatin with increasing numbers of randomly located 10 nm diameter spheres on a cylindrical surface with 11.5 nm radius, corresponding to chicken erythrocyte chromatin. The number of spheres/11 nm is 5 (A), 6 (B), 7 (C) and 8 (D). E is similar to B, but with more detailed representations of the nucleosomal DNA. A gaussian distribution with a mean value of 90 degrees and a standard deviation of 10 degrees was taken for the angle between the axis of the nucleosomal DNA superhelix and the fibre axis

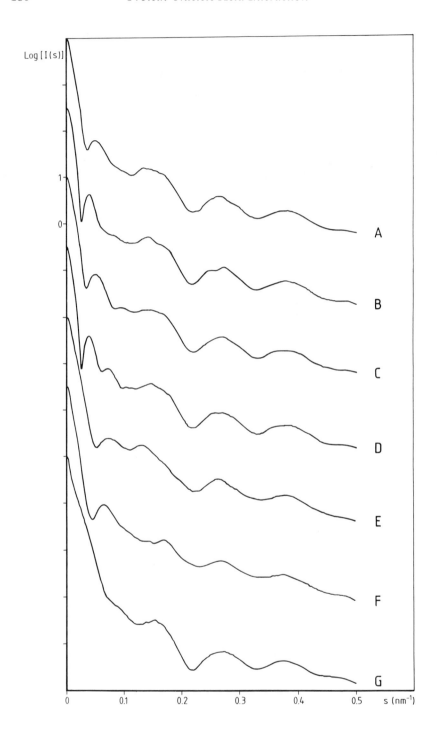

hydrodynamic measurements. The problem has been addressed by several groups using electric dichroism (for an introduction, see Frédéricq and Houssier, 1973). This requires orienting chromatin using intense, short electric field pulses and measuring the changes in absorbance for light polarized in the direction perpendicular ($A\perp$) or parallel ($A\|$) to the field. The measurements can be done only on solutions with low conductances. The absorbance of polarized light is proportional to the average square of the projection of the transition moment of the chromophore on the polarization axis. Dichroism measurements thus allow determination of the orientation of the chromophores relative to the axis of the field. In the case of chromatin, the chromophores are the DNA bases, but bound ligands like ethidium bromide or methylene blue (Kubista *et al.*, 1985) or covalently bound labels such as psoralens (Sen *et al.*, 1986) can also be used. The reduced dichroism is the ratio $\rho = \Delta A/A = (A\| - A\perp)/A$, where A is the total absorbance in the absence of field. A good check that the electric field does not distort the structure is that $A = \frac{1}{3}(A\| + 2A\perp)$. In the range of fields that do not distort the structure, full orientation is usually not achieved and $\rho = \phi\rho_\infty$, where ϕ is the orientation function $0 \leqslant \phi \leqslant 1$ and ρ_∞ the intrinsic reduced dichroism. The latter value can be obtained only by extrapolation of the reduced dichroism to infinite field in a ρ versus $1/E$ or ρ versus $1/E^2$ plot. These extrapolation procedures are not entirely satisfactory and complicate the comparison of experimental results.

For the purpose of interpretation, it is convenient to express the total reduced dichroism at complete orientation as the weight average of the contributions of the linker and the nucleosomal DNA:

$$\rho_\infty = \left[\frac{\text{repeat} - 166}{\text{repeat}}\right] \cdot \rho_{\infty, L} + \left[\frac{166}{\text{repeat}}\right]\rho_{\infty, N} \tag{8.4}$$

The intrinsic reduced dichroism of the linker DNA ($\rho_{\infty, L}$) and the nucleosome ($\rho_{\infty, N}$) for different orientations relative to the field (McGhee *et al.*, 1980) are theoretical values that are not directly accessible to experiments. For instance, the experimental value for core particles obtained by Crothers *et al.* (1978) is -0.29, which should be compared with -0.37, the theoretical value at complete orientation. The value for the linker is extrapolated from measurements on DNA. This leads to some

---

Figure 8.10 Calculated solution scattering patterns for models of condensed chicken erythrocyte chromatin. (A) is obtained as the product of the interference of a random distribution of points on a cylindrical surface and the scattering of an isolated nucleosome. (C) is obtained from the model in Figure 8.9E. (E) corresponds to the same model, but with a uniform density simulating the linker DNA in the central part of the fibre. (B, D and F) are similar patterns for a model of sea urchin chromatin with a radius of the cylindrical surface of 14.5 nm. (G) is the pattern calculated from the product of interference obtained from the coordinates of the nucleosomes given by Subirana *et al.* (1985) and the scattering of an isolated nucleosome

uncertainties about the intrinsic dichroism values of the linker DNA and the nucleosomes.

The lack of agreement between the experimental values of the reduced dichroism obtained in different laboratories is, however, more disturbing. For chicken erythrocyte chromatin, the values given by McGhee *et al.* (1983) are negative throughout the range of condensation. Although qualitatively similar results have been reported by Yabuki *et al.* (1982), there are very significant differences in the actual values. In agreement with Marquet *et al.* (1988), we found negative values at low levels of compaction and slightly positive ones for condensed samples (Koch *et al.*, 1988). In contrast, Marion *et al.* (1985a) found the birefringence to be always positive.

The values of McGhee *et al.* (1980) for the uncondensed fibre can be accounted for only by a fully extended 'beads on a string' structure. The latter is, however, incompatible with the observed relaxation times, which are ten times smaller than expected (McGhee *et al.*, 1980). This inconsistency has not been resolved. The extended 'beads on a string' is also in contradiction with the mass per unit length and radius of gyration of the cross-section obtained by X-ray scattering, as well as with the radius of gyration measured by light scattering (Ausio *et al.*, 1983). This casts some doubt on the values for the condensed fibre, which are also more negative than those found by Yabuki *et al.* (1982). The latter authors had previously reported positive values for condensed fibres stabilized by cross-linking with dimethyl suberimidate (Lee *et al.*, 1981).

For neuronal chromatin which has a repeat of 168 base pairs, Allan *et al.* (1984) found negative values ranging from -0.39 to -0.11 at increasing Mg^{2+} concentrations.

The results of flow dichroism are equally difficult to interpret. Harrington (1985), using the preparative methods of McGhee *et al.* for chicken erythrocyte chromatin, found values in agreement with those obtained by these authors (McGhee *et al.*, 1983). For calf thymus chromatin, Makarov *et al.* (1983) reported positive values at salt concentrations above 2 mM NaCl. Dimitrov *et al.* (1988) found a negative reduced dichroism for chicken erythrocyte and sea urchin chromatin but a positive one for rat liver chromatin. Kubista *et al.* (1985) found negative values below 10 mM NaCl for Ehrlich ascite chromatin, which has a linker of about 20 base pairs.

Clearly, the source of the large differences in the experimental values obtained in different, or even the same, laboratories must lie in preparative or experimental conditions that are not yet fully understood.

There have been several reports concerning the angle of tilt of the nucleosomes relative to the fibre axis, assumed to coincide with the direction of the field (Lee *et al.*, 1981; Yabuki *et al.*, 1982; McGhee *et al.*, 1983; Allan *et al.*, 1984; Mitra *et al.*, 1984). Note that although the values of the angles are similar, the experimental reduced dichroism values from which they are derived are quite different in some of these studies. Regardless of the experimental values and of the assumptions underlying the formalism, this approach has a severe limitation.

The expression giving the intrinsic dichroism of an isolated nucleosome is:

$$\rho_{\infty,N} = \tfrac{3}{8}(3\cos^2\gamma - 1) \tag{8.5}$$

A similar expression is obtained for the linker DNA. γ is the angle between the field (fibre axis) and the normal to the faces of the nucleosome. In systems characterized by a distribution of orientations, this interpretation is, however, not generally valid. If one assumes, for instance, a gaussian distribution with a mean value of 90 degrees, and a standard deviation of 30 degrees, the value of the angle obtained from the average over $\cos^2\gamma$ is 63 degrees. This gives a tilt of the plane of the nucleosomes relative to the fibre axis of $(\gamma - 90)$ or 27 degrees, close to the values given by McGhee *et al.* (1980) and Yabuki *et al.* (1982) or Allan *et al.* (1984).

Since the X-ray patterns of oriented fibres (Widom and Klug, 1985) unequivocally indicate that there is a preferred orientation of the nucleosomes with the normal to their faces perpendicular to the fibre axis ($\gamma = 90$ degrees), the value of $\rho_{\infty,N}$ obtained from equation (8.5) reflects the width of this distribution rather than its mean value. Similar considerations apply to flow dichroism results (Makarov *et al.*, 1983; Kubista *et al.*, 1985). Clarification of the experimental aspects is required to decide on the validity of the detailed solenoid models of McGhee *et al.* (1983) and Butler (1984). These models, which also assume that the fibre diameter is independent of the linker length, would have to be adapted to take evidence to the contrary discussed above into account.

LOCATION OF H1

The role of H1 in maintaining the superstructure of chromatin has been described by Thoma *et al.* (1979). These observations have been confirmed by Bordas *et al.* (1986a) and Gerchman and Ramakrishnan (1987). The effective diameter of H1 (H5)-depleted chromatin is 16 nm and the mass per unit length is reduced. In the model of Figure 8.4, these effects can be accounted for by an increase of the angle θ, resulting in an extension of the structure. This is also consistent with the large increase in negative electric dichroism of chicken erythrocyte chromatin upon removal of the H1(H5) histones (Houssier *et al.*, 1981).

The exact location of H1(H5) is, however, still unknown (for reviews, see Crane-Robinson *et al.*, 1984; Thoma, 1984). It has been suggested (Thoma *et al.*, 1979) that these histones are located at the origin and end point of the two turns of DNA superhelix on the chromatosome.

Recently, Russanova *et al.* (1987) used antibodies against the globular domains of H1 and H5 to probe the structure. They found that the antigenic determinants of the H5 globular domain remained accessible at NaCl concentrations up to 80 mM, whereas the H1 domains become inaccessible above 30 mM NaCl. Ferritin-conjugated Fab fragments obtained from anti-GH5 (Dimitrov *et al.*, 1987) do not react above 20 mM NaCl. This also suggests that the H5 globular domains are located in the interior of the fibre.

KINETICS OF CONDENSATION

The kinetics of condensation has been measured by X-ray solution scattering on chicken erythrocyte chromatin (Bordas *et al.*, 1986a) and by light scattering on rat liver chromatin (Girardet and Roche, 1985) and on calf thymus chromatin (Smirnov *et al.*, 1988). Figure 8.11 illustrates the results of the X-ray experiments. When chicken erythrocyte chromatin in TE buffer (5 mM Tris HCl, 1 mM EDTA, 0.1 mM PMSF, pH 7.5) is mixed with salt to a final concentration of 80 mM, the integrated intensity used to monitor the time course of condensation remains constant and the final pattern is that of condensed chromatin. In the control experiment, where chromatin is mixed with TE buffer, the integrated intensity also remains constant but the final pattern is that of uncondensed chromatin. This means that in solution condensation takes place in less than 50 ms, the dead-time of the mixing device. Values in the range 20–50 ms were also obtained by Smirnov *et al.* (1988). Similar experiments were performed on gels (Bordas *et al.*, 1986a). There, the process of mixing, diffusion and condensation can be separated. The half-time of the condensation process which is then diffusion-limited is of the order of seconds.

Condensation thus appears to be a very rapid shift of a continuous equilibrium rather than a structural transition from a '10 nm' filament to a '30 nm' fibre. The most important unsolved question concerning the mechanism of this fast process is whether it mainly operates by a reduction of the distance (d) between successive nucleosomes, as implied by the solenoid model, or by a reduction of the angle θ, as implied by models in which the linker DNA runs across the centre of the fibre.

SOLUBILITY PROPERTIES

The solubility properties of chromatin limit the range of conditions where valid conclusions can be drawn from scattering and hydrodynamic measurements. Condensation and aggregation followed by precipitation

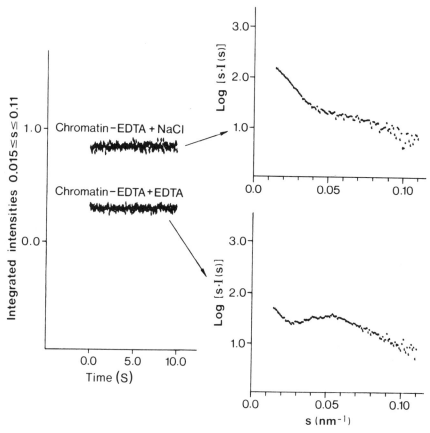

Figure 8.11   Results of X-ray stopped-flow experiments on chicken erythrocyte chromatin (A_{260} = 70) in TE buffer. The levels of the integrated intensities are constant but different in the two experiments. The final pattern at the top is characteristic of condensed chromatin, whereas in the control experiment (bottom) it is that of uncondensed chromatin

are, however, linked. To understand these phenomena, two aspects have to be considered: the intrinsic solubility properties of chromatin and the influence of various cations on solubility.

INTRINSIC SOLUBILITY PROPERTIES OF CHROMATIN

Independently of the source of material, there appear to be two distinct classes of chromatin fractions with very similar composition, but differing in their solubility properties. These fractions are called S, for soluble, chromatin, which precipitates at salt concentrations around 150 mM, and I, for insoluble, chromatin, precipitating in 60–70 mM NaCl. Ausio *et al.* (1986) also define E, for EDTA extractable, chromatin that is even less soluble. S and I chromatin fractions were found in chicken erythrocyte (Ruiz-Carillo *et al.*, 1980), calf thymus (Hollandt *et al.*, 1979) and rat liver chromatin (Brust and Harbers, 1981). The physico-chemical properties were also investigated by Fulmer and Bloomfield (1981, 1982), Komaiko and Felsenfeld (1985), Ausio *et al.* (1986), Notbohm (1986b) and Brust (1986).

Komaiko and Felsenfeld (1985) reported that the H5 content of I chromatin is higher than that of S chromatin. This was confirmed by Notbohm (1986a), who also found that condensation monitored by the reduction of the band at $(20 \text{ nm})^{-1}$ in the X-ray scattering patterns occurs at lower salt concentrations in I chromatin. These results contrast with those of the careful analysis of Ausio *et al.* (1986) on fractionated samples. In this case, it was found that if there is any difference then the H5 content is higher in S chromatin. The protein/DNA ratio of S chromatin was found to be systematically higher by a few per cent than that of I chromatin, but this can be accounted for by the presence of a small fraction of oligonucleosomes in S chromatin with about half the average protein/DNA ratio. Surprisingly, sedimentation measurements did not give any indication about a greater compaction of I chromatin, compared with S chromatin at lower NaCl concentration. The only difference between I chromatin and S chromatin was revealed by the melting profiles showing that the fraction of DNA in the linker region decreases from 29 per cent in S chromatin to 13 per cent in I chromatin. Since the repeat lengths are identical in the two fractions, this suggests that they are assembled differently, more DNA being bound to the histone core in I chromatin than in S chromatin. A detailed discussion of this aspect and of its possible physiological implications can be found in Ausio *et al.* (1986) and references therein.

These studies illustrate the subtle differences that have to be taken into account. Some of them are certainly influenced by the details of the preparative procedures, since the relative yield of S and I chromatin depends, for instance, on temperature, concentration, digestion times etc.

The fact that competing polyanions can solubilize chromatin (Lasters *et al.*, 1981) suggests that part of the differences may originate from the presence of tightly bound cations (e.g. polyamines). Other observations reinforce this suspicion. One is that sea urchin chromatin, which has a very long linker, and would thus be expected to be more soluble, already precipitates at about 60 mM NaCl. The second is that when polyamines are used in some of the buffers during chromatin extraction, positive electric birefringence (Chauvin *et al.*, 1985) and flow dichroism (Tjerneld *et al.*, 1982) are obtained, even in the absence of salt. Other groups only find positive birefringence at salt concentrations where chromatin is condensed. Despite some progress, the case of I and S chromatin and its possible physiological significance remains open.

EFFECT OF CATIONS ON CONDENSATION AND SOLUBILITY

Most investigations on the effect of cations were carried out on I chromatin. The interaction of long chicken erythrocyte chromatin fragments with monovalent cations has been investigated by light scattering (Ausio *et al.*, 1984), sedimentation (Ausio *et al.*, 1984), X-ray scattering (Bordas *et al.*, 1986a; Widom, 1986; Greulich *et al.*, 1987; Koch *et al.*, 1987a) and electron microscopy (Widom, 1986). In all cases, it was found that all monovalent cations induce compaction in a similar range of concentration (0–80 mM), followed by precipitation at higher concentrations. No differences were found in the properties of the alkali chlorides (LiCl, NaCl, KCl and CsCl) with regard to compaction (Koch *et al.*, 1987a), but precipitation by these salts was not investigated in detail. The dependence of the apparent mass per unit length on the nature and concentration of these cations (Koch *et al.*, 1987a) suggests that the number of counterions interacting with chromatin is the same for all cations. In this respect, the behaviour of chromatin is similar to that of DNA (Luzzati *et al.*, 1967). Monovalent organic cations commonly used in buffers such as Tris (Tris[hydroxymethyl]aminomethane) have the same effect as salt, as illustrated in Figure 8.12. In contrast, ammonium chloride compacts chromatin more efficiently than the other monovalent cations and in particular the alkylammonium cations, probably as a result of hydrogen bonding (Koch *et al.*, 1988).

Similar trends are found for other types of chromatin, but the salt concentrations at which precipitation occurs are lower in rat liver chromatin (Koch *et al.*, 1987a) and sea urchin chromatin (Widom *et al.*, 1985). Note, however, that there is no relationship between the repeat length and the salt concentration at which precipitation occurs. This is related to the problem of chromatin solubility mentioned above.

Divalent cations

The effects of divalent cations have been studied on chicken erythrocyte chromatin by Borochov *et al.* (1984) and Koch *et al.* (1987a). It was found that Mg^{2+} is one of the least efficient divalent cations in inducing compaction. Typical results are illustrated in Figure 8.13. The systematic studies provide clear evidence for the selectivity of the divalent cations but not for the factors defining their relative efficiency. With cations such as Cu, it is probable that base binding, a tendency that increases in the order $Mg^{2+} < Co^{2+} < Ni^{2+} < Mn^{2+} < Zn^{2+} < Cd^{2+} < Cu^{2+}$ (Barton and Lippard, 1978) becomes important. Zn is the most efficient divalent cation. Other factors, such as the ability of cations to induce aggregation, are discussed below. The higher efficiency of copper in compacting and aggregating chromatin was also detected by electric dichroism relaxation rates

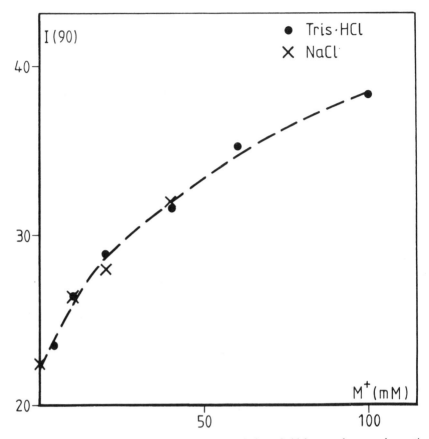

Figure 8.12   Light scattering at 90 degrees of a solution of chicken erythrocyte chromatin ($A_{260} = 20$) as a function of Tris·HCl or NaCl concentration

measurements (Sen and Crothers, 1986), but this method revealed no differences between Co, Zn, Mn and Mg.

The trends observed in solution are also applicable to gels (Staron, 1985), a situation more related to concentrations found *in vivo*. Magnesium produces a volume reduction of the gels that can be reversed with the chelator *o*-phenanthroline, whereas with calcium the effect is irreversible.

Mixtures of NaCl and MgCl$_2$

Borochov *et al.* (1984) found that the solubility of chicken erythrocyte chromatin in the presence of MgCl$_2$ is increased at higher NaCl concentration, so that higher compaction is achieved before aggregation sets in. This conclusion is based on the change of the apparent diffusion coefficient measured by light scattering and on measurements of the chromatin

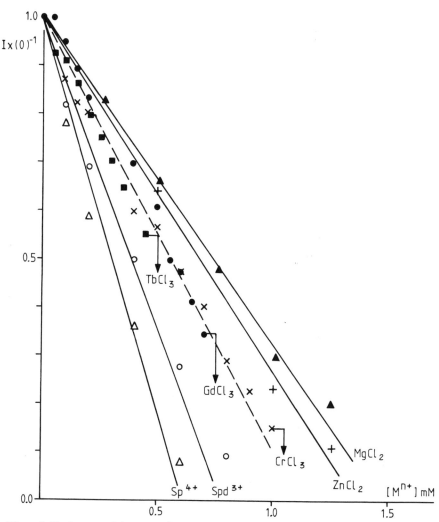

Figure 8.13 Inverse of the apparent mass/unit length, indicating progressive compaction of chicken erythrocyte chromatin ($A_{260} = 60$) as a function of cation concentration. The arrows indicate the onset of precipitation (Koch *et al.*, 1988)

concentration in the supernatant after centrifugation. The results are illustrated in Figure 8.14. Note that the apparent diffusion coefficient increases with salt concentration at any given $MgCl_2$ concentration. Addition of NaCl shifts the onset of precipitation to higher $MgCl_2$ concentrations, the major effect occurring below 5 mM NaCl. In these experiments the buffer contained 1 mM Tris.

Monovalent and divalent cations are thus synergetic with regard to compaction, but monovalent cations prevent aggregation. Preliminary

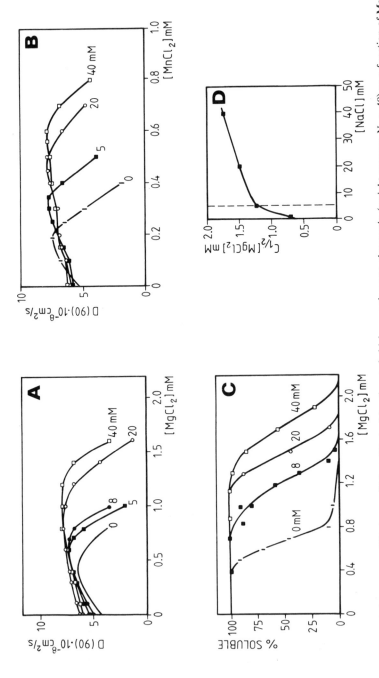

Figure 8.14   (A) Apparent diffusion coefficient (D_{90}) of fractionated chicken erythrocyte chromatin (weight average $N_w = 48$) as a function of $MgCl_2$ concentration at various NaCl concentrations (20 °C, $A_{260} = 0.8$, 1 mM Tris-HCl, pH 8.0). (B) Similar results for a fraction with $N_w = 33$, against $MnCl_2$ concentration under the same experimental conditions. (C) Solubility of chicken erythrocyte chromatin ($A_{260} = 0.8$) at 4 °C against $MgCl_2$ concentration at various NaCl concentrations, measured as a fraction of the initial absorbance in the supernatant after centrifugation at $12000 \times g$. (D) $MgCl_2$ concentration corresponding to 50 per cent solubility ($C_{1/2}$) in the previous graph as a function of NaCl concentration. (After Borochov *et al.*, 1984)

X-ray scattering measurements of the mass per unit length shown in Figure 8.15 are compatible with these results and their interpretation. These results do not cover the range below 5 mM monovalent cation concentration. Matters may not be so simple as indicated by similar results for $MnCl_2$ in Figure 8.14. There, the apparent diffusion coefficients display a more complex behaviour. Higher compaction is achieved before precipitation at increasing salt concentration, but the fact that the curves cross each other can be interpreted as resulting from an antagonistic effect of the monovalent cations.

Electron microscopy also conveys a different picture. Azorin *et al.* (1982) found that the morphology of the fibres depends on whether condensation is induced by monovalent or divalent cations. Addition of 10–50 mM NaCl at 0.5 mM $MgCl_2$ results in apparently less compact fibres than in the absence of salt. At 100 mM NaCl, the fibres are again compact. On the basis of the 'phase diagram' in Figure 8.16, obtained by X-ray scattering and electron microscopy, and of sedimentation studies on chicken erythrocyte chromatin in buffers containing NaCl and $MgCl_2$, Widom (1986) reached the conclusion that monovalent cations have a dual effect. At concentrations above 45 mM, they would act to fold chromatin into 30 nm filaments. At concentrations below this, when other multivalent cations are present and cause compaction, they would act competitively to unfold or destabilize the 30 nm filament. This means, for instance, that a significantly larger $MgCl_2$ concentration is required at 25 mM NaCl than at 0.2 mM NaCl to obtain the same sedimentation coefficient. It is not clear whether a similar difference exists between 5 mM NaCl and 25 mM NaCl.

The 'phase diagram' given by Widom (1986) merges results of X-ray scattering obtained at chromatin concentrations above 10 mg/ml and of electron microscopy corresponding to concentrations of 1 μg/ml. This may lead to difficulties, since there is dependence between the chromatin and the $MgCl_2$ concentrations required to achieve compaction and precipitation (Koch *et al.*, 1988). Nevertheless, the 'phase diagram' indicates that above 5 mM monovalent cations there is little effect of the NaCl concentration on the $MgCl_2$ concentrations that bring about compaction and precipitation. The boundaries of the 'phase diagram' are defined by visual classification of the appearance of the electron micrographs. It should, however, be taken into account that compaction is a continuous process, as also confirmed by the X-ray scattering and sedimentation data of Widom (1986).

At very low salt concentration, however, polyelectrolytes like chromatin expand (see, for instance, Tanford (1961)), as indicated by the 20 per cent increase in the value of the radius of gyration between 5 mM NaCl and 1 mM NaCl (Ausio *et al.*, 1984).

Whether the properties of chromatin can be accounted for by cation competition, as foreseen by the theories of Manning (1978) and Record *et*

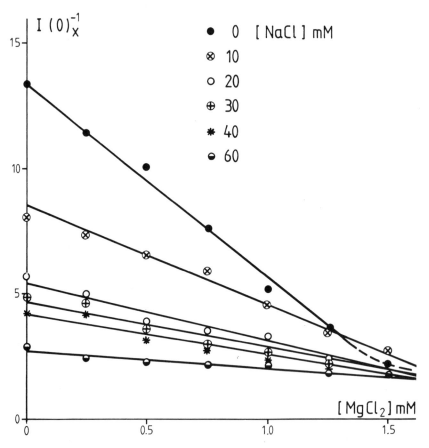

Figure 8.15 Inverse of the apparent mass/unit length of chicken erythrocyte chromatin ($A_{260} = 60$) in different mixtures of $MgCl_2$ and NaCl

al. (1978), is thus still an open question, given the fragmentary data available. These theories are certainly applicable to linear polyelectrolytes like DNA at low ionic strength (Widom and Baldwin, 1980), but it is unclear whether they also apply to chromatin.

Once more, the discrepancies between the experimental results of different groups illustrate the importance of the details of the experimental conditions in comparing results. Some of the questions could probably be clarified by careful studies in the 0–5 mM NaCl concentration range. One practical difficulty with those poorly buffered samples is to ensure that the pH does not decrease, thus leading to an alteration of the structure.

The matter has a physico-chemical interest, but it seems unlikely that cation competition effects on compaction, if they occur, play any role at physiological concentrations. An enhancement of solubility due to mixtures of cations, however, could be more significant.

Multivalent cations

The multivalent cations compact and aggregate chromatin at even lower concentrations than the divalent ones, although the differences with the most efficient divalent cations are small. Results are illustrated in Figure 8.12. The most striking feature is that the lanthanides appear to cross-link the fibres, rather than compact them, and lead easily to precipitation.

Little is known about possible competition effects between multivalent and monovalent cations. Widom (1986) observed competitive effects by sedimentation and electron microscopy. Sen and Crothers (1986), using electric dichroism, found synergetic effects. In preliminary experiments on chicken erythrocyte chromatin ($A_{260} = 60$) mixtures of NaCl (0–60 mM) with $CrCl_3$ (0–1.25 mM) and of $MgCl_2$ (0–1.5 mM) and $CrCl_3$ (0–1.25 mM) in Tris buffer (5 mM Tris HCl, 0.1 mM PMSF, pH 7.5), we found that in this range of concentration, cations had synergistic effects and that precipitation occurred at lower $CrCl_3$ concentrations.

It is evident that the understanding of many aspects of the compaction of chromatin remain very superficial. The process of compaction by cations monitored by light or X-ray scattering, as well as sedimentation, appears to be continuous and to be inseparable from aggregation, although influenced by different factors. Perhaps not surprisingly, the continuous nature of this process is less obvious in electron microscopy. There are, however, indications that, like most polyelectrolytes, chromatin has an expanded structure in absence of salt.

Most information concerning the interactions between cations and chromatin, and in particular counterion condensation, is extrapolated from observations on DNA. Counterion condensation to a polyelectrolyte in the sense defined by Manning (1978) occurs if, in an environment containing only the single counterion species, the charge fraction of the polyelectro-lyte remains constant over a broad range of concentration. At low salt concentrations, the local counterion concentration in the vicinity of the polyelectrolyte can be several orders of magnitude higher than the bulk concentration. These counterions are not tightly bound, i.e. they are not localized and dehydrated. In the case of chromatin compaction, the neutralized charge fraction must increase from an estimated 70 per cent neutralized by histones and monovalent cations to 100 per cent (Sen and Crothers, 1986), over a very narrow range of divalent cation concentration. Although most of the effects are clearly due to unspecific charge neutra-lization, in some cases true ion binding may also play a role. In this respect, an interesting observation, on the increase of conductivity of a chicken erythrocyte chromatin solution in 1 mM cacodylate buffer upon addition of $MgCl_2$, illustrated in Figure 8.17, was made by Marquet *et al.* (1988). In absence of $MgCl_2$, the conductivity of the solution is lower than that of the buffer. This is followed by an increase in conductivity paralleling that of

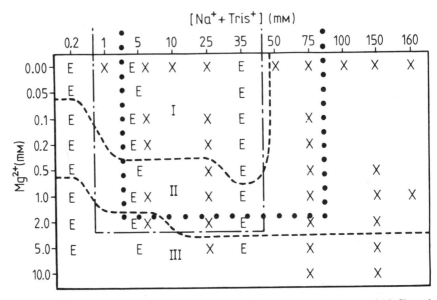

Figure 8.16 'Phase diagram' for chicken erythrocyte chromatin in mixtures of MgCl$_2$ and NaCl obtained by electron microscopy (E) and X-ray scattering (X). In region I, chromatin is unfolded as judged from electron micrographs; in region II, it is folded in the characteristic 30 nm filament; and in region III, it is aggregated (after Widom, 1986). The boundaries of the regions investigated by light scattering (Borochov *et al.*, 1984) (–·–·) and X-ray scattering (see Figure 8.13) (● ● ● ●) are indicated

the buffer. In the region corresponding to condensation (0.5–0.7 mM MgCl$_2$), there is a plateau, indicating that there is co-operative uptake of ions. From the length of the plateau that occurs well before the onset of precipitation, it was estimated that 2.75 Mg^{2+} were bound per phosphate group. At higher Mg concentrations, there is a further release of counterions accompanied by a significant increase of the slope. In calf thymus chromatin, the conductivity increase was monotonous. Such effects can be detected only in dilute solutions, where compaction and aggregation can be more easily separated. Little is known about the thermodynamics of cation-induced compaction. The process is not influenced by temperature, indicating that there is no change in enthalpy. Sen and Crothers (1986) have also attempted to calculate the entropy change associated with the binding of different cations. This contribution to the free energy change is nearly constant for monovalent, divalent and multivalent cations. This also means that the entropic contributions associated, for instance, with the folding, the release of water and the formation of hydrophobic interactions must also be constant or compensate each other. Neutralization of charge is certainly the most important factor affecting compaction and aggregation, but this can be achieved in different ways and leads to structural differences. Specific effects, such as base binding in the case of Cu or hydrogen bonding and spatial correlation of charges as in the polyamines, are, however, more likely to provide the structural, thermodynamic and kinetic selectivity characterizing cellular control mechanisms.

The balance between compaction and aggregation depends also on the ability of cations to form cross-links leading to aggregation. A theoretical approach to this problem seems difficult, but a number of observations can be rationalized on the basis of the rules given by Tam and Williams (1985). For instance, the ability to form many bonds of irregular geometry makes calcium a good cross-linking agent, whereas magnesium is not. Similarly, Na, K, and the lanthanides have few restrictions on ligand geometry, whereas Mg tends to have octahedral coordination. The direct consequence of this property of large cations is that procedures like negative staining in electron microscopy are likely to distort the structures. These properties may also explain the difference in the appearance of the fibres in electron microscopy and in the X-ray scattering patterns, depending on the nature of the cations. Magnesium tends to give smooth, more compact fibres than Na (e.g. Widom, 1986), whereas with terbium the fibres are more loosely packed (Marquet *et al.*, 1988).

A more complete explanation would also have to consider the role of competing anions (Lasters *et al.*, 1981; Sen and Crothers, 1986), and of such parameters as pH on the behaviour of chromatin.

The crucial problem is, of course, how chromatin unfolds under physiological conditions. *In vivo*, this process seems to involve removal of the histones, which neutralize about 50 per cent of the charge on the DNA. This should also be accompanied by an increase in solubility of the chromatin.

A GALLERY OF MODELS

Uncondensed chromatin

There is substantial consensus that at low salt concentrations (1 mM < [NaCl] < 20 mM), chromatin with linkers in the range 30–75 base pairs forms a (three-dimensional), zig-zag structure as proposed by Thoma *et al.* (1979). It is possible that a more extended structure exists in the absence of cations, but not at NaCl concentrations above 1 mM NaCl.

X-ray scattering indicates that in solutions and gels, the distance between nucleosomes is regular and that the linker is extended. At low resolution, the fibres have an effective diameter of 25–40 nm, depending on the linker length and a mass/unit length close to 1 nucleosome/11 nm. Their superstructure is maintained by the H1(H5) histones. Description of this structure as a '10 nm' filament has led during the last decade to controversies that can be resolved by taking into account the low resolution of hydrodynamic and scattering methods. Given the unexplained

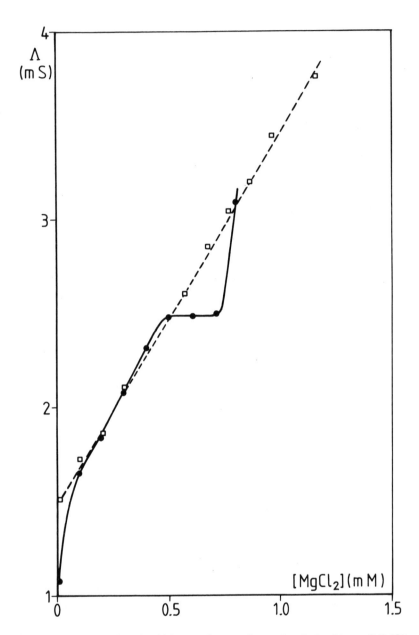

Figure 8.17   Conductivity of a chicken erythrocyte chromatin solution ($A_{260} = 0.5$) (●) and of the corresponding buffer (□) as a function of $MgCl_2$ concentration (after Marquet *et al.*, 1988)

discrepancies between electric dichroism results from different laboratories, the orientation of the nucleosomes relative to the fibre axis is still
uncertain.

Condensed chromatin

There is a large consensus on the physico-chemical parameters of condensed chromatin of the most common source – chicken erythrocytes. The
results of electron microscopy, hydrodynamic methods and scattering
techniques are all consistent. In contrast, there is no general agreement on
the fibre diameter and mass per unit length for chromatin from other
sources.

Not surprisingly, in the absence of any direct structural evidence, there is
no consensus on the path of the linker DNA. This has led to several
proposals (for a picture of these models, see Sayers, 1988).

The solenoid (Finch and Klug, 1976)

The solenoid model implies some kind of folding of the linker DNA, so
that successive nucleosomes in the unfolded chain become adjacent in
condensed chromatin. Extensions of this model, defining the path of the
linker DNA, proposed by McGhee *et al.* (1983) and Butler (1984), are
based on electric dichroism measurements that are not firmly established.

The solenoid model does not require a dependence of the fibre diameter
on the linker length, but does not exclude it. It is not clear, however, how it
applies to chromatin with a very short linker. Sedimentation measurements
on neuronal chromatin (Pearson *et al.*, 1983) indicate that there is a
structural transition at increasing salt concentration. Accurate values of the
mass per unit length and of the fibre diameter that could prove the
formation of a fibre with 6 nucleosomes/11 nm are not yet available.

Another difficulty encountered with the solenoid model is to explain the
results of some of the DNAse digestion experiments. This led Staynov
(1983) and Burgoyne (1985) to propose different models.

The triple-helix model of Makarov *et al.* (1985)

This model is based on flow dichroism measurements that are at variance
with the electric and flow dichroism of most other groups. The orientation
of the nucleosomes in this model is inconsistent with the patterns of
oriented chromatin fibres (Widom and Klug, 1985).

The double-helical crossed-linker model of Williams *et al.* (1986)

This is the most detailed model hitherto proposed. Successive nucleosomes
in the zig-zag structure of uncondensed chromatin are connected by the

linker DNA across the centre of the condensed fibre in a non-adjacent manner. The term 'crossed-linker' was introduced by Williams *et al.* (1986) to define a category of models sharing this property that also includes the two following models. Crossed-linker models imply a dependence of the fibre diameter on the linker length. This dependence has been established by electron microscopy and X-ray scattering. The experimental evidence for a left-handed double helix rests on the analysis of Fourier transforms of electron micrographs of stretches of fibre, about 50–100 nm long (for comparison, the models in Figure 8.9 correspond to a fibre length of 100 nm). These results do not provide direct evidence for the path of the linker. The helical symmetry that is detected should correspond to the arrangement of the equivalent nucleosomes (i.e. they should have the same colour in the original models). Because of the limited sampling and selection inherent in electron microscopy, it is not yet clear how representative the proposed double helical arrangement is.

The accordion model of Bordas *et al.* (1986b)

The model is based on static and time-resolved X-ray scattering. Essentially, it describes the condensation process as the collapse of the three-dimensional zig-zag structure into a short-range-order fibrous structure in which the linker crosses the central part of the fibre.

The X-ray scattering data on which this model is based do not provide any direct evidence for the path of the linker. The model predicts a dependence of fibre diameter on linker length, as observed. For chicken erythrocyte chromatin, the mass per unit length and fibre diameter are identical to those of the solenoid or to the double helical crossed-linker model.

For chromatins with a very short linker (e.g. neuronal chromatin), this model predicts an outer diameter of about 25 nm for the fibres, which would not change much with increasing salt concentration. This diameter is compatible with the electron microscopic observations of Rattner *et al.* (1982), Pearson *et al.* (1983) and Allan *et al.* (1984).

Models like the two previous ones, in which the linker runs across the centre of the fibre, also provide a basis for the explanation of the nuclease digestion experiments. The following two models were specifically designed to interpret these experiments.

The non-sequential model of Staynov (1983)

This model proposes a non-sequential but regular arrangement of nucleosomes (Staynov, 1983; Crane-Robinson *et al.*, 1984). The linked DNA criss-crosses the inside of the fibre. In its original version, the model imposes geometric constraints that appear too stringent, but the basic idea is in fact incorporated in the two previous models. A more detailed analysis of DNAse cutting patterns for chromatin from various sources would be very useful, to decide not only between the solenoid and crossed-linker models but also between the different crossed-linker models.

The back-to-back zig-zag model of Burgoyne (1985)

This model also aims mainly at explaining the disomal repeat in DNAse I digestion patterns of nuclei (Burgoyne and Skinner, 1981; Pospelov *et al.*, 1982; Drinkwater *et al.*, 1987). This repeat, determined by field inversion gel electrophoresis, is coherent to 16 N (Davis and Burgoyne, 1988). These results probably represent the strongest evidence against regularly coiled models hitherto available. In its original form, the model does not suffice to explain the physico-chemical data, and in particular the continuous nature of the compaction process. Backfolding or lateral aggregation of filaments would also be expected to lead to characteristic interference bands in the X-ray scattering patterns, which are not observed. Nevertheless, the experimental observations on which this model rests require an explanation, which has not yet satisfactorily been given by other models.

The twisted-ribbon model of Worcel *et al.* (1981)

The model also stresses the importance of the basic zig-zag structure in the formation of the condensed fibre. Like the following model, it requires a rather regular helical structure, for which there is hitherto no convincing evidence.

The twisted-ribbon model of Woodcock *et al.* (1984)

Based on electron microscopic observations, this model is similar to the previous one. The zig-zag is folded into a helix to form a hollow cylinder with an inner diameter of 10 nm and an outer diameter of 32 nm. In its most condensed form, chicken erythrocyte chromatin would have 12 nucleosomes/11 nm. Assuming a volume of 540 nm^3 for the nucleosome, this leads to a packing ratio of 0.81, which is larger than the values in dense crystalline solids. Moreover, since the linker DNA runs from one layer of nucleosomes to the next along the fibre axis, the condensed fibre should have a large negative value for the reduced dichroism, in contradiction to all experimental observations.

The superbead model of Renz *et al.* (1977)

This model is based on the observation of globular supranucleosomal particles at physiological salt concentrations (see also Strätling *et al.*, 1978).

All of the models presented above, with the exception of the twisted ribbons, could easily be modified to form superbead structures. The model in fact already describes a higher level of organization of chromatin, where regions devoid of nucleosomes, which could correspond to transcriptionally active chromatin, are interspersed with condensed ones.

CONCLUSION

The structural problems that remain to be solved at the level of resolution discussed above are those concerning the path of the DNA, the location of H1 and the arrangement of nucleosomes in short repeat chromatin. In particular, experiments will have to be designed that allow an unequivocal decision between the solenoid model and crossed-linker models with similar structural parameters (mass/unit length and fibre diameter). The models that offer a satisfactory agreement with the physicochemical data will also have to be more thoroughly tested against the results of biochemical approaches.

Resolving some of the remaining paradoxes will lead to a better understanding, not only of the structure and condensation mechanism of chromatin but also of the relationship between the results of various experimental approaches.

REFERENCES

Allan, J., Rau, D. C., Harborne, N. and Gould, H. (1984). Higher order structure in short repeat length chromatin. *J. Cell Biol.*, **98**, 1320–1327

Annunziato, A. T. and Seale, R. L. (1983). Chromatin replication, reconstitution and assembly. *Molec. Cell. Biochem.*, **55**, 99–112

Ausio, J., Borochov, N., Kam, Z., Reich, M., Seger, D. and Eisenberg, H. (1983). In Helene, C. (ed.), *Structure, Dynamics, Interactions and Evolution of Biological Macromolecules*, Reidel Publishing Co., New York, 89–100

Ausio, J., Borochov, N., Seger, D. and Eisenberg, H. (1984). Interaction of chromatin with NaCl and MgCl$_2$: solubility and binding studies. Transition to and characterization of the higher order structure. *J. Mol. Biol.*, **177**, 373–398

Ausio, J., Sasi, R. and Fasman, G. D. (1986). Biochemical and physiochemical characterization of chromatin fractions with different degrees of solubility isolated from chicken erythrocyte nuclei. *Biochemistry*, **25**, 1981–1988

Azorin, F., Perez-Grau, L. and Subirana, J. A. (1982). Supranucleosomal organization of chromatin. *Chromosoma*, **85**, 251–260

Barton, J. K. and Lippard, J. (1978). Heavy metal interaction with nucleic acids. In Spiro, T. G. (ed.), *Nucleic Acid–Metal Ions Interactions*, John Wiley and Sons, New York, 31–113

Baudy, P. and Bram, S. (1979). Neutron scattering on nuclei. *Nucleic Acid Res.*, **6**, 1721–1729

Bordas, J., Perez-Grau, L., Koch, M. H. J., Nave, C. and Vega, M. C. (1986a). The superstructure of chromatin and its condensation mechanism. I: Synchrotron radiation X-ray scattering results. *Eur. Biophys. J.*, **13**, 157–174

Bordas, J., Perez-Grau, L., Koch, M. H. J., Nave, C. and Vega, M. C. (1986b). The superstructure of chromatin and its condensation mechanism. II: Theoretical analysis of the X-ray scattering patterns and model calculations. *Eur. Biophys J.*, **13**, 175–186

Borochov, N., Ausio, J. and Eisenberg, H. (1984). Interaction and conformational changes of chromatin with divalent ions. *Nucleic Acid Res.*, **12**, 3089–3096

Bradbury, E. M. and Baldwin, J. P. (1986). Neutron scatter and diffraction techniques applied to nucleosome and chromatin structure. *Cell Biophysics*, **9**, 35–66

Brust, R. and Harbers, E. (1981). Structural investigations on isolated chromatin of higher-order organisation. *Eur. J. Biochem.*, **117**, 609–615

Brust, R. (1986). Soluble and insoluble rat liver chromatin is different in structure and protein composition. *Z. Naturforsch.*, **41c**, 910–916

Burgoyne, L. A. (1985). A back-to-back zig-zag model for higher order chromatin structure. *Cytobios.*, **43**, 141–147

Burgoyne, L. A. and Skinner, J. D. (1981). Chromatin superstructure, the next level after the nucleosome has an alternating character. A two-nucleosome based series is generated by probes armed with DNAse-I acting on isolated nuclei. *Biochem. Biophys. Res. Commun.*, **99**, 893–899

Butler, P. J. G. (1984). A defined structure of the 30 nm chromatin fibre which accommodates different nucleosomal repeat lengths. *EMBO J.*, **3**, 2599–2604

Cantor, C. R. and Schimmel, P. R. (1980). *Biophysical Chemistry*, Freeman and Co., San Francisco

Chauvin, F., Roux, B. and Marion, C. (1985). Higher order structure of chromatin: influence of ionic strength and proteolytic digestion on the birefringence properties of polynucleosomal fibers. *J. Biomol. Struct. Dyn.*, **2**, 805–819

Crane-Robinson, C., Staynov, D. Z. and Baldwin, J. P. (1984). Chromatin higher order structure and histone H1. *Comments Mol. Cell. Biophys.*, **2**, 219–265

Crothers, D. M., Dattagupta, N., Hogan, M., Klevan, L. and Lee, K. S. (1978). Transient electric dichroism studies of nucleosomal particles. *Biochemistry*, **17**, 4525–4532

Damaschun, H., Damaschun, G., Pospelov, V. A. and Vorob'ev, V. I. (1980). X-ray small-angle scattering study of mononucleosomes and the close packing of nucleosomes in polynucleosomes. *Molec. Biol. Rep.*, **6**, 185–191

Davis, S. J. and Burgoyne, L. A. (1988). The DNAse I generated disomal series is coherent to 16N: implications for coiling models of chromatin structure. *FEBS Lett.*, **226**, 88–90

Dimitrov, S. I., Russanova, V. R. and Pashev, I. G. (1987). The globular domain of histone H5 is internally located in the 30 nm chromatin fibre: an immunological study. *EMBO J.*, **6**, 2387–2392

Dimitrov, S. I., Smirnov, I. V. and Makarov, V. L. (1988). Optical anisotropy of chromatin: Flow linear dichroism and electric dichroism studies. *J. Biomol. Struct. Dyn.*, **5**, 1135–1148

Drinkwater, R. D., Wilson, P. R., Skinner, J. D. and Burgoyne, L. A. (1987). Chromatin structures: dissecting their patterns in nuclear digests. *Nucleic Acids Res.*, **15**, 8087–8103

Eissenberg, J. C., Cartwright, I. L., Thomas, G. H. and Elgin, S. C. R. (1985). Selected topics in chromatin structure. *Ann. Rev. Genet.*, **19**, 485–536

Felsenfeld, G. and McGhee, J. D. (1986). Structure of the 30 nm chromatin fiber. *Cell*, **44**, 375–377

Finch, J. T. and Klug, A. (1976). Solenoidal model for superstructure in chromatin. *Proc. Natl Acad. Sci. USA*, **73**, 1897–1901

Flory, P. J. (1953). *Principles of Polymer Chemistry*, Cornell University Press, Ithaca

Frédéricq, E. and Houssier, C. (1973). *Electric Dichroism and Electric Birefringence*, Clarendon Press, Oxford

Fulmer, A. W. and Bloomfield, V. A. (1981). Chicken erythrocyte nucleus contains two classes of chromatin that differ in micrococcal nuclease susceptibility and solubility at physiological ionic strength. *Proc. Natl Acad. Sci. USA*, **78**, 5968–5972

Fulmer, A. W. and Bloomfield, V. A. (1982). Higher order folding of two different classes of chromatin isolated from chicken erythrocyte nuclei. A light scattering study. *Biochemistry*, **21**, 985–992

Gerchman, S. E. and Ramakrishnan, V. (1987). Chromatin higher order structure studied by neutron scattering and scanning transmission electron microscopy. *Proc. Natl Acad. Sci. USA*, **84**, 7802–7806

Giradet, J. L. and Roche, J. (1985). Dynamic changes in chromatin superstructure: A light scattering study. *Studia Biophys.*, **107**, 13–21

Greulich, K. O., Wachtel, E., Ausio, J., Seger, D. and Eisenberg, H. (1987). Transition of chromatin from the '10 nm' lower order structure, to the '30 nm' higher order structure, as followed by small angle X-ray scattering. *J. Mol. Biol.*, **193**, 709–721

Harrington, R. E. (1985). Optical model studies of the salt induced 10–30 nm fiber transition in chromatin. *Biochemistry*, **24**, 2011–2021

Hollandt, H., Notbohm, H., Riedel, F. and Harbers, E. (1979). Studies of the structure of isolated chromatin in three different solvents. *Nucleic Acid Res.*, **6**, 2017–2027

Houssier, C., Lasters, I., Muyldermans, S. and Wyns, L. (1981). Influence of histones H1/H5 on the DNA coiling in the nucleosome–electric dichroism and birefringence study. *Int. J. Biol. Macromol.*, **3**, 370–376

Igo-Kemenes, T., Hörz, W. and Zachau, H. G. (1982). Chromatin. *Ann. Rev. Biochem.*, **51**, 89–121

Koch, M. H. J., Vega, M. C., Sayers, Z. and Michon, A. M. (1987a). The superstructure of chromatin and its condensation mechanism. III: Effect of monovalent and divalent cations. X-ray solution scattering and hydrodynamic studies. *Eur. Biophys. J.*, **14**, 307–319

Koch, M. H. J., Sayers, Z., Vega, M. C. and Michon, A. M. (1987b). The superstructure of chromatin and its condensation mechanism. IV: Enzymatic digestion, thermal denaturation, effect of netropsin and distamycin. *Eur. Biophys. J.*, **15**, 133–140

Koch, M. H. J., Sayers, Z., Michon, A. M., Marquet, R., Houssier, C. and Willführ, J. (1988). The superstructure of chromatin and its condensation mechanism. V: Effect of linker length, condensation by multivalent cations, solubility and electric dichroism properties. *Eur. Biophys. J.*, **16**, 177–185

Komaiko, W. and Felsenfeld, G. (1985). Solubility and structure of domains of chicken erythrocyte chromatin containing transcriptionally competent and inactive genes. *Biochemistry*, **24**, 1186–1193

Kubista, M., Hard, T., Nielsen, P. E. and Norden, B. (1985). Structural transitions of chromatin at low salt concentrations: a flow linear dichroism study. *Biochemistry*, **24**, 6336–6342

Langmore, J. P. and Schutt, C. (1980). The higher order structure of chicken erythrocyte chromosomes in vivo. *Nature*, **288**, 620–622

Langmore, J. P. and Paulson, J. R. (1983). Low angle X-ray diffraction studies of chromatin structure in vivo and in isolated nuclei and metaphase chromosomes. *J. Cell Biol.*, **96**, 1120–1131

Lasters, I., Muyldermans, S., Wyns, L. and Hamers, R. (1981). Differences in rearrangements of H1 and H5 in chicken erythrocyte chromatin. *Biochemistry*, **20**, 1104–1110

Lasters, I., Wyns, L., Muyldermans, S., Baldwin, J., Poland, G. A. and Nave, C. (1985). Scatter analysis of discrete-sized chromatin fragments favours a cylindrical organization. *Eur. J. Biochem.*, **151**, 283–289

Lee, K. S., Mandelkern, M. and Crothers, D. M. (1981). Solution structural studies of chromatin fibers. *Biochemistry*, **20**, 1438–1445

Luzzati, V., Masson, F., Mathis, A. and Saludjian, P. (1967). Etude par diffusion centrale des Rayons X, des polyélectrolytes rigides en solution, cas des sels de Li, Na et Cs du DNA. *Biopolymers*, **5**, 491–508

Makarov, V. L., Dimitrov, S. I., Tsaneva, I. R. and Pashev, I. G. (1983). Salt-induced conformational transitions in chromatin. *Eur. J. Biochem.*, **133**, 491–497

Makarov, V., Dimitrov, S., Smirnov, V. and Pashev, I. (1985). A triple helix for the structure of chromatin fiber. *FEBS Lett.*, **181**, 357–361

Mandelbrot, B. M. (1983). *The Fractal Geometry of Nature*, W. H. Freeman, New York

Manning, G. S. (1978). The molecular theory of polyelectrolyte solutions with applications to the electrostatic properties of polynucleotides. *Quart. Rev. Biophys.*, **11**, 179–246

Marion, C., Martinage, A., Tirard, A., Roux, B., Daune, M. and Mazen, A. (1985a). Histone phosphorylation in native chromatin induces local structural changes as probed by electric birefringence. *J. Mol. Biol.*, **186**, 367–379

Marion, C., Hesse-Bezot, C., Bezot, P., Marion, M-J., Roux, B. and Bernengo, J. C. (1985b). The effect of histone H1 on the compaction of oligo nucleosomes – a quasi-elastic light scattering study. *Biophys. Chem.*, **22**, 53–64

Marquet, R., Colson, P., Matton, A. M., Houssier, C., Thiry, M. and Goessens, G. (1988). Comparative study of the condensation of chicken erythrocyte and calf thymus chromatins by di- and multivalent cations. *J. Biol. Struct. Dyn.*, **5**, 839–857

McGhee, J. D., Rau, D. C., Charney, E. and Felsenfeld, G. (1980). Orientation of the nucleosome within the higher order structure of chromatin. *Cell*, **22**, 87–96

McGhee, J. D., Nickol, J. M., Felsenfeld, G. and Rau, D. C. (1983). Higher order structure of chromatin. Orientation of nucleosomes within the 30 nm chromatin solenoid is independent of species and spacer length. *Cell*, **33**, 831–841

Mitra, S., Sen, D. and Crothers, D. M. (1984). Orientation of nucleosomes and linker DNA in calf thymus chromatin determined by photochemical dichroism. *Nature*, **308**, 247–250

Nagl, W. (1986). Chromatin organization and the control of gene activity. *Internat. Rev. Cytol.*, **94**, 21–56

Nelson, W. G., Pienta, K. J., Barrack, E. R. and Coffey, D. S. (1986). The role of the nuclear matrix in the organization and function of DNA. *Ann. Rev. Biophys. Biophys. Chem.*, **15**, 457–475

Notbohm, H. (1986a). Small angle scattering of cell nuclei. *Eur. Biophys. J.*, **13**, 367–372

Notbohm, H. (1986b). Comparative studies on the structure of soluble and insoluble chromatin from chicken erythrocytes. *Int. J. Biol. Macromol.*, **8**, 114–120

Olins, A. L. and Olins, D. E. (1974). Spheroid chromatin units (*v*-bodies). *Science*, **183**, 330–332

Oster, G. and Riley, D. P. (1952). Scattering from cylindrically symmetric systems. *Acta Cryst.*, **5**, 272–276

Oudet, R., Gross-Bellard, M. and Chambon, P. (1975). Electron microscopic and biochemical evidence that chromatin structure is a repeating unit. *Cell*, **4**, 281–300

Pearson, E. C., Butler, P. J. G. and Thomas, J. O. (1983). Higher-order structure of nucleosome oligomers from short repeat chromatin *EMBO J.*, **2**, 1367–1372

Pederson, D. S., Thoma, F. and Simpson, R. T. (1986). Core particle, fiber and transcriptionally active chromatin structure. *Ann. Rev. Cell Biol.*, **2**, 117–147

Perez-Grau, L., Azorin, F. and Subirana, J. A. (1982). Aggregation of mono- and dinucleosomes into chromatin-like fibers. *Chromosoma*, **87**, 437–445

Perez-Grau, L., Bordas, J. and Koch, M. H. J. (1984). Synchrotron radiation X-ray scattering study on solutions and gels. *Nucleic Acids Res.*, **12**, 2987–2995

Porod, G. (1982). General theory. In Glatter, O. and Kratky, O. (eds), *Small Angle X-ray Scattering*, Academic Press, New York, 17–53

Pospelov, V. A., Svetlikova, S. B. and Vorob'ev, V. I. (1982). Nucleodisome – a repeating unit of chromatin structure. *Studia biophys.*, **87**, 119–120

Prunell, A. and Kornberg, R. D. (1982). Variable center to center distance of nucleosomes in chromatin. *J. Mol. Biol.*, **153**, 515–523

Ramsay-Shaw, B. and Schmitz, K. S. (1976). Quasielastic light scattering by biopolymers. Conformation of chromatin multimers. *Biochem. Biophys. Res. Comm.*, **73**, 224–232

Rattner, J. B., Saunders, C., Davie, J. R. and Hamkalo, B. A. (1982). Ultrastructural organization of yeast chromatin. *J. Cell Biol.*, **92**, 217–222

Renz, M., Nehls, P. and Hozier, J. (1977). Involvement of histone H1 in the organization of the chromosome fibre. *Proc. Natl Acad. Sci. USA*, **74**, 1879–1883

Record, M. T., Jr, Anderson, C. F. and Lohman, T. M. (1978). Thermodynamic analysis of ion effects on the binding and conformational equilibria of proteins and nucleic acids: the roles of ion association or release, screening and ion effects on water activity. *Quart. Rev. Biophys.*, **11**, 103–178

Richmond, T. J., Finch, J. T., Rushton, B., Rhodes, D. and Klug, A. (1984). Structure of the nucleosome core particle at 7 Å resolution. *Nature*, **311**, 532–537

Ruiz-Carillo, A., Puigdomenech, P., Eder, G. and Lurz, R. (1980). Stability and reversibility of higher order structure of interphase chromatin: continuity of deoxyribonucleic acid is not required for maintenance of folded structure. *Biochemistry*, **19**, 2544–2554

Russanova, V. R., Dimitrov, S. I., Makarov, V. L. and Pashev, I. G. (1987). Accessibility of the globular domain of histones H1 and H5 to antibodies upon folding of chromatin. *Eur. J. Biochem.*, **167**, 321–326

Sayers, Z. (1988). Synchrotron X-ray scattering studies of the chromatin fibre structure. In E. Mandelkow (ed.), *Topics in Current Chemistry*, **145**, Springer Verlag, Heidelberg, 203–232

Schmitz, K. S. and Ramsay-Shaw, B. (1977). Chromatin conformation, a systematic analysis of helical parameters from hydrodynamic data. *Biopolymers*, **16**, 2619–2633

Sen, D. and Crothers, D. M. (1986). Condensation of chromatin, role of multivalent cations. *Biochemistry*, **25**, 1495–1503

Sen, D., Mitra, S. and Crothers, D. M. (1986). Higher order structure of chromatin, evidence from photochemically detected linear dichroism. *Biochemistry*, **25**, 3441–3447

Smirnov, I. V., Dimitrov, S. I. and Makarov, V. L. (1988). NaCl-induced chromatin condensation. Application of static light scattering at 90° and stopped-flow technique. *J. Biol. Struct. Dyn.*, **5**, 1127–1134

Staron, K. (1985). Condensation of the chromatin gel by cations. *Biochim. Biophys. Acta*, **825**, 289–298

Staynov, D. Z. (1983). Possible nucleosome arrangements in the higher order structure of chromatin. *Int. J. Biol. Macromol.*, **5**, 3–9

Strätling, W. H., Muller, U. and Zentgraf, H. (1978). The higher order repeat structure of chromatin is built up of globular particles containing eight nucleosomes. *Exp. Cell Res.*, **117**, 301–311

Suau, P., Bradbury, E. M. and Baldwin, J. P. (1979). Higher-order structures of chromatin in solution. *Eur. J. Biochem.*, **97**, 593–602

Subirana, J. A., Munoz-Guerra, S., Aymami, J., Radermacher, M. and Frank, J. (1985). The layered organization of nucleosomes in 30 nm fibers. *Chromosoma*, **91**, 377–390

Tam, S. C. and Williams, R. J. P. (1985). Electrostatics and biological systems. *Structure and Bonding*, **63**, 103–151

Tanford, C. (1961). *Physical Chemistry of Macromolecules*, John Wiley and Sons, New York

Thoma, F., Koller, T. and Klug, A. (1979). Involvement of histone H1 in the organization of the nucleosome and of the salt dependent superstructures of chromatin. *J. Cell Biol.*, **83**, 403–407

Thomas, J. O. (1984). The higher order structure of chromatin and histone H1. *J. Cell Sci.*, *Suppl. I*, 1–20

Thomas, J. O., Rees, C. and Butler, P. J. G. (1986). Salt-induced folding of sea urchin sperm chromatin. *Eur. J. Biochem.*, **154**, 343–348

Tjerneld, F., Nordén, B. and Wallin, H. (1982). Chromatin structure studied by linear dichroism at different salt concentrations. *Biopolym.*, **21**, 343–358

Vainshtein, B. K. (1966). *Diffraction of X-rays by Chain Molecules*, Elsevier Publishing Co., Amsterdam

Walker, P. R. and Sikorska, M. (1987). Chromatin structure, evidence that the 30 nm fiber is a helical coil with 12 nucleosomes/turn. *J. Biol. Chem.*, **262**, 12223–12227

Widom, J. (1986). Physicochemical studies of the folding of the 100 Å nucleosome filament into the 300 Å filament. *J. Mol. Biol.*, **190**, 411–424

Widom, J. and Baldwin, R. L. (1980). Cation-induced toroidal condensation of DNA. Studies with $Co^{3+}(NH_3)_6$. *J. Mol. Biol.*, **144**, 431–453

Widom, J. and Klug, A. (1985). Structure of the 300 Å filament, X-ray diffraction from oriented samples. *Cell*, **43**, 207–213

Widom, J., Finch, J. T. and Thomas, J. O. (1985). Higher-order structure of long repeat chromatin. *EMBO J.*, **4**, 3189–3194

Williams, S. P., Athey, B. D., Muglia, L. J., Schappe, R., Gough, A. J. and Langmore, J. P. (1986). Chromatin fibres are left-handed double helices with diameter and mass per unit length that depend on linker length. *Biophys. J.*, **49**, 233–250

Woodcock, C. L. F., Frado, L. L. Y. and Rattner, J. B. (1984). The higher order structure of chromatin, evidence for a helical ribbon arrangement. *J. Cell Biol.*, **99**, 42–52

Worcel, A. S., Strogatz, S. and Riley, D. (1981). Structure of chromatin and the linking number of DNA. *Proc. Natl Acad. Sci. USA*, **78**, 1461–1465

Wu, R. S., Panusz, H. T., Hatch, C. L. and Bonner, W. M. (1986). Histones and their modifications. *CRC Crit. Rev. Biochem.*, **20**, 201–263

Yabuki, H., Dattagupta, N. and Crothers, D. M. (1982). Orientation of nucleosomes in the thirty-nanometer chromatin fiber. *Biochemistry*, **21**, 5015–5020

Yaniv, M. and Cereghini, S. (1986). Structure of transcriptionally active chromatin. *CRC Crit. Rev. Biochem.*, **21**, 1–26

Zentgraf, H. and Franke, W. W. (1984). Differences of supranucleosomal organization in different kinds of chromatin. Cell type-specific globular subunits containing different numbers of nucleosomes. *J. Cell Biol.*, **99**, 272–286

Index